Stem Cells - Laboratory and Clinical Research

Human Mesenchymal Stem Cells

STEM CELLS - LABORATORY AND CLINICAL RESEARCH

Additional books and e-books in this series can be found on Nova's website under the Series tab.

STEM CELLS - LABORATORY AND CLINICAL RESEARCH

HUMAN MESENCHYMAL STEM CELLS

MITCHELL KHAN
EDITOR

Copyright © 2021 by Nova Science Publishers, Inc.

All rights reserved. No part of this book may be reproduced, stored in a retrieval system or transmitted in any form or by any means: electronic, electrostatic, magnetic, tape, mechanical photocopying, recording or otherwise without the written permission of the Publisher.

We have partnered with Copyright Clearance Center to make it easy for you to obtain permissions to reuse content from this publication. Simply navigate to this publication's page on Nova's website and locate the "Get Permission" button below the title description. This button is linked directly to the title's permission page on copyright.com. Alternatively, you can visit copyright.com and search by title, ISBN, or ISSN.

For further questions about using the service on copyright.com, please contact:
Copyright Clearance Center
Phone: +1-(978) 750-8400   Fax: +1-(978) 750-4470   E-mail: info@copyright.com

NOTICE TO THE READER

The Publisher has taken reasonable care in the preparation of this book, but makes no expressed or implied warranty of any kind and assumes no responsibility for any errors or omissions. No liability is assumed for incidental or consequential damages in connection with or arising out of information contained in this book. The Publisher shall not be liable for any special, consequential, or exemplary damages resulting, in whole or in part, from the readers' use of, or reliance upon, this material. Any parts of this book based on government reports are so indicated and copyright is claimed for those parts to the extent applicable to compilations of such works.

Independent verification should be sought for any data, advice or recommendations contained in this book. In addition, no responsibility is assumed by the Publisher for any injury and/or damage to persons or property arising from any methods, products, instructions, ideas or otherwise contained in this publication.

This publication is designed to provide accurate and authoritative information with regard to the subject matter covered herein. It is sold with the clear understanding that the Publisher is not engaged in rendering legal or any other professional services. If legal or any other expert assistance is required, the services of a competent person should be sought. FROM A DECLARATION OF PARTICIPANTS JOINTLY ADOPTED BY A COMMITTEE OF THE AMERICAN BAR ASSOCIATION AND A COMMITTEE OF PUBLISHERS.

Additional color graphics may be available in the e-book version of this book.

Library of Congress Cataloging-in-Publication Data

ISBN: 978-1-53619-155-4

Published by Nova Science Publishers, Inc. † New York

Contents

Preface   vii

Chapter 1   Mesenchymal Stem Cells for the Treatment of COVID-19: Why and When They Should Be Used?   1
Hakan Darici, Eda Sun, Duygu Koyuncu Irmak and Erdal Karaöz

Chapter 2   Activated Mesenchymal Stem Cells for Stroke Repair   53
Ravi Prakash, Santosh Kumar Yadav, Abu Junaid Siddiqui, Neha Kumari, Mohsin Ali Khan and Syed Shadab Raza

Chapter 3   Starvation Rations: The Therapeutic Potential of Ketone Bodies for Stem Cell Function   101
Mary Board

Index   129

PREFACE

In Chapter 1, the COVID-19 pandemic and the damage mechanisms on the cellular level which can be ameliorated with the cellular therapies is thoroughly evaluated. Previous and ongoing stem cell clinical trial data from diseases with similar symptoms is gathered. All this accumulated data and current clinical trial results indicate that the cellular therapies could be the most effective treatment option for COVID-19 patients to ameliorate the damaged tissues and save lives. In Chapter 2, the authors examine activated mesenchymal stem cells for stroke repair. Stem Cell treatment has shown recovery in animal models of stroke, indicating an improved regenerative and repair potential. Though stem cells are still being used in clinical trials, there is no evidence that they enhance recovery in ischemic stroke patients. Nevertheless, the multipotent mesenchymal stem has widely been explored for stroke recovery. An 'Activated MSC' as a therapeutic alternative to tackling ischemic stroke is proposed, thereby the activation of MSCs by cytokines, growth factors, hypoxia, pharmacological drugs, etc., could be a novel approach to improving stroke patients' responses to receiving MSCs. In Chapter 3, the potential benefits of in vitro culture of therapeutic stem cells in the presence of HB along with the ketogenic diet, whereby higher physiological concentrations of ketone bodies can be achieved in vivo, as an adjuvant to stem cell transplantation is assessed.

Chapter 1 - COVID-19 is a pandemic caused by severe acute respiratory syndrome coronavirus-2 (SARS-CoV-2), which caused deaths of more than 300.000 people around the world within the first few months of 2020. SARS-CoV-2 uses ACE2 receptors to infect respiratory system cells and may cause pneumonia and severe lung damage. The virus can also spread other organs rapidly via ACE2 expressing endothelial cells and cause coagulopathy, and further damage to the organs. Another and probably more harmful effect of the virus is the overreaction of the immune system leading to hyperinflammation causing multiple organ failure and death. Therefore COVID-19 can be considered as a viral infection causing auto-immune disorders. Currently, no vaccine or effective pharmacological treatment established for the disease. On the other hand, Mesenchymal Stem Cells (MSCs) possess anti-inflammatory and immune-regulatory effects along with their regenerative abilities. In this article, the authors thoroughly evaluate the COVID-19 pandemic and the damage mechanisms on the cellular level which can be ameliorated with the cellular therapies. The authors also gathered previous and ongoing stem cell clinical trial data from diseases with similar symptoms. All these accumulated data and current clinical trial results indicate that the cellular therapies could be the most effective treatment option for COVID-19 patients to ameliorate the damaged tissues and save lives.

Chapter 2 - Cerebral stroke is a severe health concern. Stem Cell treatment has shown recovery in animal models of stroke, indicating an improved regenerative and repair potential. Though stem cells are still being used in clinical trials, there is no evidence that they enhance recovery in ischemic stroke patients. Nevertheless, the multipotent mesenchymal stem has widely been explored for stroke recovery. Low in vivo MSC survival, intrinsic differences between MSC, sources and donor variability, and variability of culturing protocols have been described as few limitations in the field. Strikingly, post-transplantation MSCs could be regulated by the locally regulated environment, indicating that restorative variability could be managed by selecting a priming regimen to rectify a given pathology precisely. In stem cell treatment, a new area of research is the preconditioning of cells before transplantation. As no effective treatment for stroke recovery is available, and owing to the fact that MSCs could be customized, the authors propose 'Activated MSC' as a therapeutic alternative to tackling ischemic stroke. Therefore, the activation of MSCs by cytokines, growth factors, hypoxia, pharmacological drugs, etc., could be a novel approach to improving stroke patients' responses to receiving MSCs.

Chapter 3 - Ketone bodies, produced primarily by the liver and, to a lesser extent, by intestinal epithelial cells, have traditionally been regarded as a fuel of starvation which allows the body to integrate stores of fat with those of carbohydrate and to deploy glucose-sparing mechanisms. More recently, renewed scrutiny of these starvation substrates has been stimulated by observations of the neuroprotective and anti-cancer effects of the ketogenic diet and the anti-aging consequences of caloric restriction. Ketone bodies, D-3-hydroxybutyrate (HB) and acetoacetate, have multifarious effects in addition to their roles as oxidative, energy-yielding substrates. Signalling through G-protein coupled receptors serves to integrate fat and carbohydrate metabolism, promoting starvation-appropriate substrate-selection to achieve a glucose-sparing effect. Consumption of a ketone body substrate also reduces production of reactive oxygen species (ROS) in many cell-types. Reduced rates of ROS-production may drive the preference shown by human mesenchymal stem cells for a ketone body substrate, with consumption rates being up to 35-fold higher than those of glucose, accompanied by a 45-fold suppression of ROS-production. Selection of HB as an oxidative substrate, along with other properties of stem cells, such as the residence in hypoxic niches within the body, may reduce the oxidative damage, including mutation and accelerated apoptosis, that accompany oxidative stress. HB has multifarious effects on gene expression including inhibitory activity towards Class I and Class IIa histone deacetylases and an ability to cause hydroxybutyrylation of histone proteins. As a result, expression of genes which reduce oxidative stress is promoted. Expression and activity of cell cycle regulators are also affected, contributing to observed suppression of proliferation and maintenance of stemness in intestinal stem cells and influencing the processes of differentiation and apoptosis. Such observations regarding the impact of ketone bodies on stem cells have implications for stem cell therapies which have been promoted for neurodegenerative disorders and for myocardial damage, among other conditions. This review assesses the potential benefits of *in vitro* culture of therapeutic stem cells in the presence of HB along with the ketogenic diet, whereby higher physiological concentrations of ketone bodies can be achieved *in vivo*, as an adjuvant to stem cell transplantation. Thus, the ketogenic diet, long recognised for its ability to ameliorate refractory epilepsy and its favourable impact on neurological function and more recently proposed as an anti-cancer therapy, may have renewed applications as an accompaniment to stem cell therapy.

In: Human Mesenchymal Stem Cells
Editor: Mitchell Khan
ISBN: 978-1-53619-155-4
© 2021 Nova Science Publishers, Inc.

Chapter 1

MESENCHYMAL STEM CELLS FOR THE TREATMENT OF COVID-19: WHY AND WHEN THEY SHOULD BE USED?

Hakan Darici[1,2,3,*], *Eda Sun*[2], *Duygu Koyuncu Irmak*[1,2] *and Erdal Karaöz*[1,2,3,4]

[1]Istinye University, Faculty of Medicine,
Dept. of Histology & Embryology, Istanbul, Turkey
[2]Istinye University, Stem Cell, and Tissue Engineering R&D Center,
Istanbul, Turkey
[3]Istinye University, 3D Bioprinting Design & Prototyping R&D Center,
Istanbul, Turkey
[4]Liv Hospital, Stem Cell, and Regenerative Therapies Center
(LivMedCell), Istanbul, Turkey

---

[*] Corresponding Author's E-mail: hdarici@istinye.edu.tr.

ABSTRACT

COVID-19 is a pandemic caused by severe acute respiratory syndrome coronavirus-2 (SARS-CoV-2), which caused deaths of more than 300.000 people around the world within the first few months of 2020. SARS-CoV-2 uses ACE2 receptors to infect respiratory system cells and may cause pneumonia and severe lung damage. The virus can also spread other organs rapidly via ACE2 expressing endothelial cells and cause coagulopathy, and further damage to the organs. Another and probably more harmful effect of the virus is the overreaction of the immune system leading to hyperinflammation causing multiple organ failure and death. Therefore COVID-19 can be considered as a viral infection causing auto-immune disorders. Currently, no vaccine or effective pharmacological treatment established for the disease. On the other hand, Mesenchymal Stem Cells (MSCs) possess anti-inflammatory and immune-regulatory effects along with their regenerative abilities. In this article, we thoroughly evaluate the COVID-19 pandemic and the damage mechanisms on the cellular level which can be ameliorated with the cellular therapies. We also gathered previous and ongoing stem cell clinical trial data from diseases with similar symptoms. All these accumulated data and current clinical trial results indicate that the cellular therapies could be the most effective treatment option for COVID-19 patients to ameliorate the damaged tissues and save lives.

1. INTRODUCTION

Coronavirus disease 2019 or COVID-19 (WHO 2020a) was declared as a global pandemic by the World Health Organization (WHO) on 11 March 2020 (WHO 2020b). Since the cause of the pandemic determined as a novel coronavirus type, it is initially named as 2019-nCoV but following further analyses, the name was corrected as Severe acute respiratory syndrome coronavirus 2 or SARS-CoV-2 (Gorbalenya et al. 2020). Number of the infected people going up to 40 million around the world, while around one quarter in living in the USA only. The disease took the lives of more than one million people so far (Dong, Du, and Gardner 2020). It is the second pandemic, which the WHO member states acted together, after influenza A (H1N1), or with more common name, swine flu pandemic, which caused 100.000-400.000 deaths during two years in 2009 and 2010 (Dawood et al. 2012; Viboud and Simonsen 2012). Two major outbreaks of coronavirus related diseases caused severe symptoms and deaths of thousands in the past (Habibzadeh and Stoneman 2020). Around 8.000 people from 26 countries died during 2002-2003 Severe Acute Respiratory Syndrome Coronavirus (SARS-CoV) and 858 people from 2012 the Middle East Respiratory Syndrome Coronavirus (MERS-CoV) (Bruce Aylward (WHO); Wannian Liang (PRC) 2020). However current coronavirus pandemic already passed the maximum estimated numbers H1N1 pandemic related deaths within months instead of years.

Currently, 2240 clinical trials were registered to the NIH and only 49 of them (~2.2%) are using stem cells. On the other hand, about half of the roundly 8000 stem cell clinical trials were focused on diseases like pneumonia, infections, ARDS, liver-kidney-respiratory-cardiovascular-nervous system related injuries, inflammation, and immune system (www.clinicaltrials.gov). As it will be evaluated thoroughly in this chapter, COVID-19 is related to all these disorders. Therefore, we have focused on the prognosis and related molecular mechanisms of the COVID-19 related symptoms with causes and the effects, and related them to therapeutic applications and benefits of stem cells, especially, mesenchymal stem cells (MSCs), which were found to be overlapping with COVID-19 symptoms. We also collected the completed or ongoing clinical stem cell trials related to COVID-19 along with our own ongoing clinical trial results.

2. SARS-CoV-2 AND COVID-19

2.1. Virus Structure

Along with SARS-CoV, and MERS-CoV, already circulating HKU1, NL63, OC43 and 229E strains of *Coronaviridae* family make up to 2,47% of the annual cases of flu, usually with mild symptoms, according to a 2018 study (S. fen Zhang et al. 2018). *Coronaviridae* contains enveloped RNA viruses with a diameter ranging from 60 to 140 nm. Spike like surface proteins, shown under the electron microscope gives them their specific crown-like appearance and hence their naming as coronaviruses (Singhal 2020). SARS-CoV-2 is a zoonotic virus closely related to the SARS-like coronaviruses observed in Rhinolophus (horseshoe bats), with 98.7% nucleotide similarity in bats (Lai et al. 2020), indicating the disease is probably transmitted from bats, which are consumed as food in Wuhan, China, the designated origin point of the disease (Singhal 2020). One of the first strains of the virus isolated in the Wuhan seafood market has a 29,9 kb long genome which contains several open reading frames (ORFs), which has started and stop codons for protein translation. Other than 16 non-structural proteins, structural proteins for matrix (M), nucleocapsid (N), envelope (E), spike (S), and additional accessory proteins are coded. Spikes (S proteins) allows the virus to attach its receptor and initiate virus-cell membrane fusion. M protein allows the formation of the envelope and later transport of necessary molecules for viral integrity. E and N proteins possibly work against the antiviral responses of the host cell (Guo et al. 2020).

2.2. Transmission

Upon the discovery of the first patients infected with the virus in Wuhan were market dealers working with live or freshly slaughtered animals, pointing a zoonotic virus, passing from animals to humans. However, following new cases among the healthcare providers showed that the disease can also spread from patients to healthy individuals, which proved human-to-human transmission. Later research revealed the structure, genome, and similarity of the virus with the SARS-CoV (Habibzadeh and Stoneman 2020).

Transmission of the virus usually starts with inhalation of airborne droplets or contacting viral particles with facial mucosae like nose, eyes, and mouth. Studies show that the airborne droplets may suspend in the air, circulate within the room, or through air recycling channels for several hours (Kampf et al. 2020; Guo et al. 2020). After the first estimated animal to human transmission in the seafood market of Wuhan, the virus started to spread between humans. Case related studies showed the transmission occurs mostly with close contact (Guo et al. 2020). The R_0 value of SARS-CoV-2 was initially calculated as 2,68 in China (Wu et al. 2020).

2.3. Infection with SARS-CoV-2

While other strains of α-coronaviruses affect the upper respiratory tract, SARS-CoV-2 targets the lower respiratory system, especially bronchioles and alveoli (Guo et al. 2020). Upon entry to the respiratory system, the virus uses angiotensin-converting enzyme receptor 2 (ACE2) to attach and enter target cells. Genome comparing analyses of the novel coronavirus showed that a difference with SARS-CoV, in 27 amino acids, which changes the structure of the spike (S) protein. But none of these changes were on the receptor-binding domain (RBD), where the virus attaches the ACE receptor (Ge et al. 2013; Wu et al. 2020). Probably due to these mutations, S protein-ACE2 binding efficiency of the SARS-CoV-2 virus was found 10 to 20-fold higher than SARS-CoV (Song et al. 2018). S protein needs to be primed before attachment and transmembrane protease, serine 2 (TMPRSS2) enzyme found to be necessary for the preparation of S protein prior to binding to ACE2 receptors. Therefore, TMPRSS2 inhibitors could be a target for COVID-19 treatment which was proved its effectiveness with a previously approved protease inhibitor, targeting TMPRSS2 blocked the entry of the virus to the cells (Hoffmann et al. 2020).

ACE2 is a transmembrane enzyme, responsible for the cleavage of the Angiotensin II, and lower blood pressure. ACE2 can be found in many organs such as, hearth, kidney, small intestine, and brain but the most profound location is the lungs, which is the starting point of the COVID-19 symptoms. Main cell types expressing ACE2 protein are the alveolar type II (ATII) cells, enterocytes, endothelial cells, and smooth muscle cells (Turner 2015). Therefore, all these organs are primary targets of the SARS-CoV-2. However, the virus first needs to enter the circulation to reach secondary organs when inhaled as droplets. Therefore, the first target of the virus is the ACE2 positive cells in the respiratory tract and oral mucosa. A 2005 study showed the attachment of SARS-CoV happens on the apical surfaces of airway epithelial cells, rather than basolateral surfaces (Jia et al. 2005) In a recent study, using bulk RNA-seq, ACE2 expression was also found in the oral cavity, especially on the tongue epithelial cells which allow the virus to enter host cells and circulation even before reaching the lungs. The study also showed a very high expression of ACE2 on colon, gallbladder, cardiomyocytes, kidney, and many other organs (Xu et al. 2020). ACE2 and TMPRSS2 co-expression found in the sustentacular cells of the olfactory area, which makes them another target for the virus and may explain the 'loss of smell' symptom which is seen in around one-third COVID-19 patients (Sungnak, Huang, Bécavin, Berg, and Network 2020; Krishan Gupta, Sanjay Kumar Mohanty, Siddhant Kalra, Aayushi Mittal, Tripti Mishra, Jatin Ahuja, Debarka Sengupta 2020; Fodoulian et al. 2020). TMPRSS2 and ACE2 are also co-expressed in superficial conjunctiva, ciliated and secretory cells of the nasal cavity and bronchi, ATII cells, gallbladder, common bile duct and enterocytes of ileum and colon, making them easier targets for attachment (Sungnak, Huang, Bécavin, Berg, Queen, et al. 2020).

ATII cells, which are responsible for the production of surfactant and protecting surface tension, also act as stem cells for the renewal of type I alveolar cells responsible for gas exchange (Nabhan et al. 2018). Following the binding of RBD of S protein to ACE2 receptors, the virus uses other subunits of the S protein to initiate endocytosis to enter the host cell. Viral nucleocapsid (N) protein represses the cellular RNA interference system to bypass the intracellular immune system and starts the production of viral proteins and RNA without entering the cell nucleus. Matrix (M) proteins gather synthesized viral envelope proteins together and shape the curvature and structure of the virions. N protein also attaches the copied RNA molecules to the newly forming viral envelope and helps the formation. Envelope (E) proteins role in the assembly, and release of the new viruses and spread of the disease to the extracellular matrix (ECM) and other cells (Pradesh et al. 2020).

2.4. Clinical Features of COVID-19

2.4.1. Symptoms and Diagnosis

While most of the COVID-19 infected patients remain asymptomatic, some exhibit coughing and shortness of breath which are counted as major symptoms. American Centers for Disease Control and Prevention (CDC) lists symptoms as; muscle pain, headache, sore throat, fever, chills, repeated shaking with chills, and newly acquired loss of taste or smell, noting that two of these symptoms together may indicate the disease (CDC 2020). Since the symptoms mostly indistinguishable from other respiratory system infections, diagnosis of COVID-19 requires additional verification with reverse transcription-polymerase chain reaction (RT-PCR), computational tomography (CT), analysis of cytokines in blood and emergence of acute respiratory distress syndrome (ARDS) (Singhal 2020). However, most of these symptoms are observed in the later stages of the disease which may cause the delay in diagnosis and advancement of the disease. Currently, RT-PCR is used to diagnose patients, however, errors like inappropriate sampling, degradation of viral RNA before analysis, technician errors, or insufficient virus numbers for detection, mislead initial RT-PCR tests to be negative (Udugama et al. 2020). A study from China with 1014 patients, showed that RT-PCR tests were found positive for only 59% of the suspected patients while CT images showed 88% positivity. On the other hand, 97% of the RT-PCR confirmed patients were positive in CT scans, pointing that CT imaging could be a better alternative to distinguish patients (Ai et al. 2020). Generally, samples were taken from pharyngeal or nasal cavities via swabs, however, sampling via bronchoalveolar lavage found to give significantly more correct results (32% and 63% vs. 93%, respectively) (Pradesh et al. 2020; W. Wang et al. 2020).Interestingly, stool samples were generally positive with the patients for viral RNA, even after the patients ameliorated diarrhea (Xie et al. 2020), which is a rare symptom for COVID-19 but common for SARS (Habibzadeh and Stoneman 2020). Although it is not on the CDC's COVID-19 symptoms list, diarrhea is a common symptom, observed in one-third of the patients, possibly due to the replication of the viruses which may also attach and enter enterocytes due to their ACE2 receptors (Lamers et al. 2020). Another diagnosis tool for COVID-19 is the ground-glass opacities in both lungs in CT scans, indicating viral pneumonia in most patients. However, this damage usually observed after the fifth day of the infection when the symptoms like coughing or shortness of breath became prominent (Habibzadeh and Stoneman 2020; Xie et al. 2020).

2.4.2. Pathogenesis of COVID-19

According to the proposal of Siddiki and Mehra, COVID-19 patients be followed under three stages as early infection, pulmonary and hyperinflammatory phases. Most of the patients are asymptomatic or have mild symptoms like fever or cough and remain in Stage I, where their immune system fights the virus. During the second phase, pneumonia causes shortness of breath (Stage IIA) and advances to hypoxia (Stage IIB). Viral replication diminishes at the phase where the inflammatory response increases. During the third phase, the main enemy of the body is its hyper-inflammation response which causes organ damage and even death (Siddiqi and Mehra 2020).

Since the common entry path of the virus is the respiratory tract, initial damage can be seen in the bronchi and alveoli. The first hit is done by the virus itself upon the invasion of the ACE2 positive cells like ATIIs. Due to the death of the ATII cells, which has a dual function as stem cells and surfactant secretion; normal function and regeneration ability of the lungs impair, leading to ARDS. Additionally, cell death causes inflammation in alveoli and following the migration of immune cells to the damaged area (Habibzadeh and Stoneman 2020; Xie et al. 2020).

During the infection with viruses, infected cells initiate IFN-α and IFN-β secretion to warn nearby cells for viral infection and attract cytotoxic cells to eliminate the viral threat. $CD8^+$ cytotoxic T cells and NK cells migrate to the area and check the infected cells from MHC-I receptors for foreign antigens. Upon detection of viral particles, cytotoxic cells shot the infected cell with enzymes like perforin and granulysin to create gaps on the cellular membrane and direct cells to apoptosis (Dotiwala et al. 2016). Meantime, these cytotoxic cells secrete IFN-γ to attract macrophages for phagocytosis. They also secrete IL-2, which acts through autocrine signaling and further induces these cells for the following attacks. Macrophages phagocyte the cell debris, disintegrate, and present viral antigens via MHC-II receptors to $CD8^+$ cells or T_H cells. These T cells also use IL-2 to activate T_C cells and IFN-γ to attract NK cells and phagocytic cells. Other immune cells like neutrophils are also attracted to the area to fight the infection through paracrine signaling (Akdis et al. 2016).

While the innate immune system tries to fight infection, rapidly cumulating viruses, subsequent damage to the $ACE2^+$ cells and intensifying inflammatory response, leads to a massive increase of cytokines like interferon-gamma (IFN-γ), tumor necrosis factor-alpha (TNFα), granulocyte colony-stimulating factor (GCSF), monocyte chemoattractant protein 1 (MCP-1), Interferon gamma-induced protein 10 (IP10), and interleukins (IL) especially IL-1, IL-2, IL-4, IL-7, IL-10, IL-12, IL-13 (Guo et al. 2020; Singhal 2020). This overstimulation phenomenon is now known as "cytokine storm", which the terms like 'macrophage activation syndrome (MAS)' or 'secondary haemophagocytic lymphohistiocytosis' also used to describe an excessive immune response to external stimuli. Cytokine storm emerges as one of the most important factors of the mortality of the COVID-19, via causing the ARDS or extrapulmonary multiple-organ failure, and consequently death (Ye, Wang, and Mao 2020; Guo et al. 2020).

The damage done by the cytokine storm is occurring via several mechanisms and it affects the clinical outcome to ARDS or multiple organ failure (MOF) scenarios. In the ARDS scenario, increasing direct damage of the virus to alveoli is further enhanced by the inflammation. Pulmonary edema, a hallmark of inflammation causes alveoli to fill with extracellular fluid and obstructs the airways, causing shortness of breath and death in 30–45% of normal ARDS cases (Horie et al. 2018). Neutrophil activation and migration to the alveolar space during the inflammation process of COVID-19 causes another problem with cytokine storm. Activated neutrophils sacrifice themselves and throw their nuclear DNA to the extracellular space to form a network. Combining with the mucus in the airways, these network of DNA forms a unique structure, called neutrophil extracellular traps (NETs), form webs within the alveoli and catches pathogens. However, in COVID-19, an excess number of NETs can block the alveolar entry, prevent extracellular cellular fluid that is accumulating due to the inflammation to be expelled, impair the function of the alveoli and cause ARDS (Barnes et al. 2020).

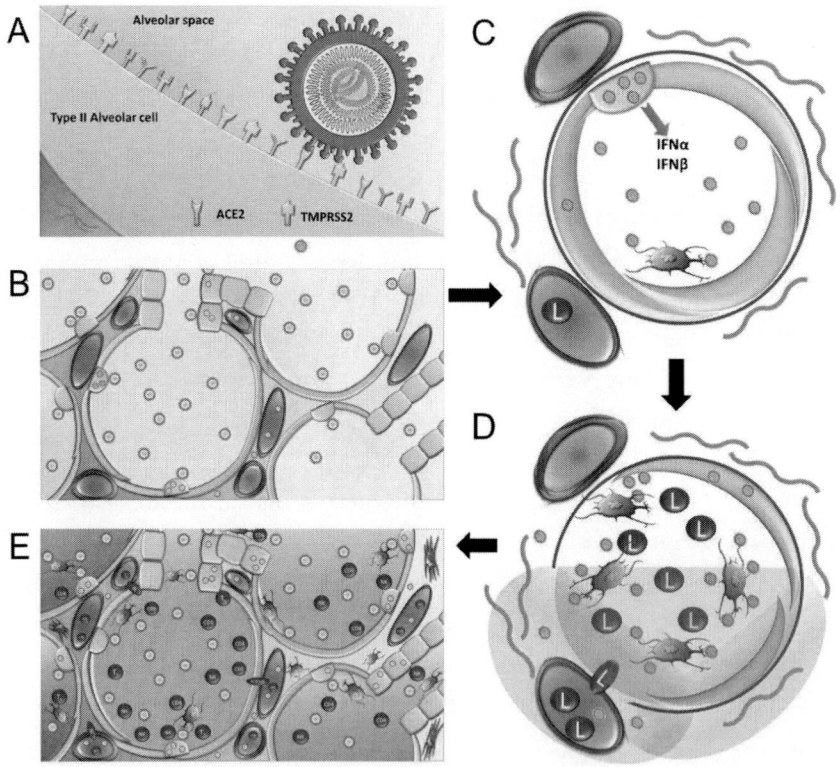

Figure 1. Effects of SARS-CoV-2 to the lungs. The virus uses ACE2 receptors on the Type II alveolar cells or bronchial epithelia to enter the cytoplasm (A), replicates and copies attack invade alveolar space (B). IFNα and IFNβ signals from the infected cells invites immune cells (C). Infection and dying cells disrupt alveolar integrity, cause inflammation, and edema (D). Spreading viruses stimulate a further immune response. The excessive immune response causes hyper-inflammation, accumulation of immune cells, and fluid within the alveolar space therefore, hindering the alveolar gas exchange function (D). (ACE2, angiotensin-converting enzyme receptor 2; IFNα and IFNβ, interferon-alpha and -beta; L, leukocyte; TMPRSS2, transmembrane protease, serine 2).

2.4.3. Extrapulmonary Damage

The initial target of the virus, the ACE2 receptors, also plays a role in the damage through its impaired function on vasodilatation. Binding of the viral S proteins to the ACE2 receptors blocks the attachment of the Angiotensin II and therefore, cleavage into inactive, angiotensin-1,7 form. Consequently, angiotensin II remains active in circulation and carries out its function to stimulate vasoconstriction (Turner 2015; Pradesh et al. 2020).

A cohort study pointed that the mortality of the disease is strongly correlated with disseminated intravascular coagulation (DIC) where 71,4% of dead patients developed DIC while only 0,6% of the survivors had the same condition (Tang, Li, et al. 2020). Increased D-dimer levels in the mortal cases also emerges as an indicator of coagulation problems (Zhou et al. 2020). Following these leads, the cause of the multi-organ damage can be tied to the activation of coagulation cascades within the blood vessels. Since endothelial cells express ACE2, they can be infected by the SARS-CoV-2. Like the alveolar cells, these endothelial cells die due to the production of new viruses causing endothelial damage (Hess, Eldahshan, and Rutkowski 2020). Elevated levels of d-dimer, elongated prothrombin time (PT), low platelet and fibrinogen levels in COVID-19 patients indicates coagulopathy, which may develop due to the endothelial damage and hyper-inflammation, also supports this phenomena (Henderson et al. 2020; Hess, Eldahshan, and Rutkowski 2020). During coagulopathy or DIC, micro-thrombosis occurs within the blood vessels around the virus-infected and damaged endothelial cells, causing the ischemia on smaller vessels and following the formation of bigger clots. Meanwhile, widespread thrombosis causes thrombocytopenia, the depletion of thrombocytes, which sequentially causes insufficient clotting and hemorrhage at the other virus damaged areas (Figure 2). The effects of these two phenomena are further increased with vasoconstriction associated with ACE2 depletion and Angiotensin II increase which consequentially may result in myocardial infarcts (MI), strokes, or organ failure and death (Hess, Eldahshan, and Rutkowski 2020).

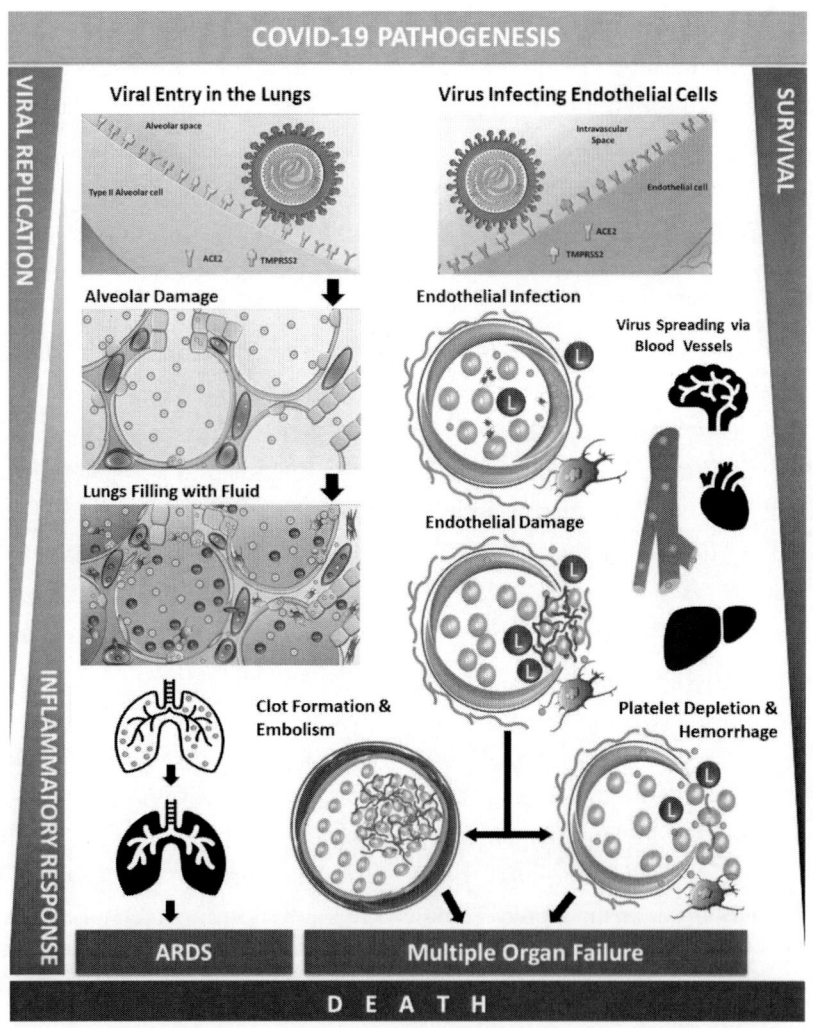

Figure 2. Schematic diagram of the COVID-19 pathogenesis starting with viral replication and advancing with alveolar and endothelial damage leading to ARDS and MOF and consequently death in worst-case scenarios. Viral entry and replication are followed by immune response and which can turn into cytokine storm. Pulmonary infection may cause alveolar damage, fluid accumulation, and ARDS. On the other hand, infection of endothelial cells may cause coagulopathy, embolism, and/or hemorrhage within organs leading to MOF. Both scenarios could be lethal and survival chance drops with developing symptoms. (ACE2, angiotensin-converting enzyme receptor 2; ARDS, acute respiratory distress syndrome; MOF, multiple organ failure; TMPRSS2, transmembrane protease, serine 2).

A case report from New York confirmed that large-vessel stroke in 5 young individuals, where vascular endothelial dysfunction and coagulopathy were thought to be the underlying reason (Oxley et al. 2020). Also, treatment effectiveness with anti-thrombogenic agents like low molecular weight heparin shows the prognosis of the disease is mostly related to coagulopathy and DIC (J. Wang et al. 2020). Infiltration of the virus into the circulation, via target cells may result with the infection of the other organs with ACE2 positive cells, like kidneys, intestines, and heart. Replication and accumulation of the virus in these organs causes inflammation, the loss of function and symptoms like kidney failure, liver failure, or MI, which may lead to permanent organ damage or death (Gheblawi et al. 2020). All these ARDS, DIC, and MOF scenarios are equally discouraging for both doctors and patients and make the nature of the COVID-19 more unpredictable and dangerous.

3. MSCs and Cytokine Storm

3.1. Properties of MSCs

Mesenchymal Stem Cells (MSCs) refer to the concept of a stem cell type, which can differentiate both *in vitro* and *in vivo* into adipogenic, osteogenic, chondrogenic, and neurogenic lineages (Karaöz et al. 2011). Although MSCs sometimes referred to multipotent stem or stromal cells, the most accepted term is still Mesenchymal Stem Cells (Le Blanc 2006; Naji et al. 2019). However, Caplan, who initially named MSCs, suggested that the term be changed into 'medical signaling cells', pointing their signaling properties and clinical potential (Caplan 2017) which we will mention thoroughly.

MSCs can be isolated from adipose tissues, bone marrow, umbilical cord (Wharton Jelly, WJ), dental pulp, periodontal ligament, even pancreatic islets (Kabatas et al. 2018; Erdal Karaoz et al. 2019). However, for clinical use, newborn umbilical cord stroma is recommended due to high differentiation capacities (El Omar et al. 2014; Okur et al. 2018). A useful attribute of MSCs is they are well tolerated when transplanted donor-to host, even though fully mismatched cells were used, indicating they are not limited with MCH compatibility (Le Blanc 2006).

Along with their proliferation and differentiation abilities, MSCs have numerous biological properties associated with the therapeutic uses, like; migration and targeting (homing), trophic, anti-apoptotic, anti-inflammatory, anti-oxidant, anti-bacterial, anti-fibrotic, chemoattractant, neuroprotective and immunosuppressant properties (Figure 3). They use various signaling molecules to carry out these effects such as:

- Vascular cell adhesion molecule (VCAM), C-X-C chemokine receptor type 4 (CXCR4) and CD44 for homing,
- Brain-derived neurotrophic factor (BDNF), hepatocyte growth factor (HGF), vascular endothelial growth factor (VEGF) for trophic effects,
- Granulocyte macrophage-colony stimulating factor (GM-CSF), HGF, VEGF, TGF-β, bFGF, and IL-6 for anti-apoptotic effects,
- Matrix metalloproteinase-2 (MMP-2), MMP-9, tissue inhibitor of metalloproteinase-1 TIMP-1, TIMP-2, Angiopoietin-1 (Ang-1) and VEGF for anti-fibrotic effects,
- Ang-1, basic fibroblast growth factor (bFGF), insulin-like growth factor (IGF-1), platelet-derived growth factor (PDGF), HGF, IL-6, and VEGF for promoting angiogenesis.

They also secrete antibacterial molecules like hepcidin, beta-defensin, and cathelicidin (LL-37) to fight infections. On the other hand, signaling roles of MSCs over the inflammation usually occur via indoleamine 2,3-dioxygenase (IDO), prostaglandin-E2 (PGE-2), IL-10, and TGF-β molecules (Karantalis and Hare 2015; Maumus, Jorgensen, and Noël 2013; Naji et al. 2019; Demircan et al. 2011). MSCs also produce extracellular matrix molecules, including fibronectin, laminin, integrins, and collagen (Le Blanc 2006).

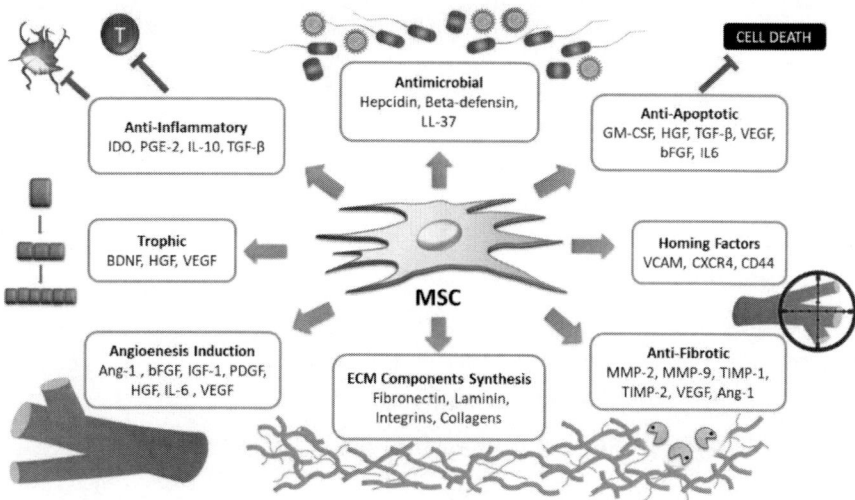

Figure 3. Signaling molecules of MSCs in tissue repair, infection, inflammation, ECM, and immune modulation.

Stem cells are particularly known with their ability to target the damaged area in the body and differentiate and replace damaged cells. However, only a small part of them shown to survive in the host organism. Studies point out the MSC treatments are effective not only via differentiating to other cells but also as through paracrine signaling (Jayaraman et al. 2013). Signaling may occur via individually secreted molecules, receptor-receptor interactions, or extracellular vesicles (EVs). EVs are thought to be the most important mediators for the clinical efficacy of MSCs, which are functional in the transport of paracrine signal molecules. Exosomes are usually 30-100 nm wide membrane covered EVs, secreted by all cell types in the body. They can contain a wide variety of proteins, mRNA, microRNA, and even DNA. Exosomes can be specifically demonstrated and isolated by their membrane receptors, CD9, CD63, CD81, and CD82 (Sarvar, Shamsasenjan, and Akbarzadehlaleh 2016). MSC exosomes contain multiple signaling molecules to carry out immunomodulatory, angiogenetic, and cell proliferation (Shabbir et al. 2015; J. Zhang et al. 2015; B. Zhang et al. 2015).

3.2. Immunomodulatory and Anti-Inflammatory Effects of MSCs

Over 2000 entries are present in PubMed for MSCs and inflammation together while around 500 of them directly mention the anti-inflammatory effects of MSCs. Similarly, around 700 articles on PubMed mention the immunomodulatory effects of MSCs. Since COVID-19 causes inflammation in the lungs, and hyperactivation of the immune system; anti-inflammatory and immunomodulatory effects of MSC therapies could be quite useful for COVID-19 treatment.

During the SARS-CoV-2 infection, an increasing amount of pro-inflammatory cytokines such as IL-1β, IL-2, IL-6, IL-8 IFNs, and TNF-α causes migration of numerous macrophages and neutrophils into the alveolar space (Fanelli et al. 2013). While the virus itself causing damage to the surfactant secreting ATII and neighboring endothelial cells, activation of cytotoxic and phagocytic cells also damage the alveolar structure. Following the fallen blood-air barriers; extracellular fluid and blood leak to the alveolar space. Migrating immune cells also fills the area and causes serious inflammation, fluid accumulation, or collapse of the alveoli (Barnes et al. 2020). These events explain the ground-glass view on the CT sections, which is an indicator of COVID-19 (X. Li et al. 2020).

MSCs possess immunomodulatory properties that point these cells to be used clinically in the treatment of various immune diseases such as graft-versus-host disease (GvHD) or multiple sclerosis (MS). These properties of MSCs include; suppression of T and B cell proliferation, B cell terminal differentiation, NK cell, and macrophage activation, dendritic cell maturation, and function (Uccelli, Moretta, and Pistoia 2008; English and Hillier 2010). MSCs perform these effects via cell-cell interactions and paracrine secretions, including exosomal contents (Prockop et al. 2010). Immunomodulatory effects have been demonstrated in the suppression of T cell proliferation through TNF-α and IFN-γ secretions in the co-culture of T cells and MSCs via both direct cell-to-cell contact and the paracrine factors such as exosomes (van den Akker et al. 2018). We also have demonstrated immunosuppressive properties of MSCs with phytohemagglutinin-stimulated T cells-MSC co-culture where anti-inflammaroty cytokines were increased and apoptotic genes were upregulated in T cells (Erdal Karaoz et al. 2017). We will evaluate these effects below on the aspects of immune cells and signaling molecules.

3.2.1. MSCs and Immune Cells

T-helper type 1 (T_h1) cells are activated by the inflammatory cytokines, which is an important step of specific immunity (Ye, Wang, and Mao 2020). Cytokines secreted by T_h1 cells like IL-2 and IFN-γ attract and recruit other T cells and macrophages and prepare cells for antiviral response (Gafa et al. 2006). Studies show that MSCs inhibit the activation effects of T_h1 cells via regulating the secretion cross-talks of IFNG and Il-1B (Uccelli, Moretta, and Pistoia 2006). The regulation of T_h1 responses is based on both the PGE2-dependent and myeloid cell-mediated mechanism (Rozenberg et al. 2016). There are two faces of PGE2 on Th1 is regulation (Luz-Crawford et al. 2012), it is not only the regulate differentiation, but also can control its proliferation. PGE2 controls Th1 proliferation via the Programmed cell death protein-1 (PD-1) pathway. The PD-1 pathway serves as a brake for T cells. In the presence of PGE2, PD-1 and PD-L1 are attached to the T cell, and Th1 proliferation is suppressed, while the immune system is also restrained.

Cytotoxic $CD8^+$ T lymphocytes (T_c cells) are one of the first cells, responding to inflammatory cytokines. MSCs' main immunoregulatory effects are based on inhibiting the T cell population (Vieira Paladino et al. 2017). In this regulatory mechanism, $CD8^+$ cells are suppressed via secretion of PGE2, IDO, and TGF-β and repressing NKG2D expression (M. Li et al. 2014). Detailed *in vitro* studies have demonstrated that MSCs decrease IFN-γ and TNF-α production while increasing anti-inflammatory IL-10, thereby limiting T cell expansion (Caplan 2007).

NK cells are another type of cytotoxic immune cells that kills infected or cancer cells, in a way similar to $CD8^+$ T cells (Rosenberg and Huang 2018). MSCs are shown to inhibit the proliferation and cytotoxic effects of IL-2 activated NK cells. Additionally, they can inhibit IFN-γ production of

NK cells, therefore have an inhibitory effect on the stimulant capacity of the NK cells, via PGE2 secretions (Uccelli, Moretta, and Pistoia 2008).

Macrophages differentiate from monocytes and may exist in two different phenotypes as pro-inflammatory M1 and anti-inflammatory M2 macrophages. M1 macrophages can be activated in response to IFN-γ, colony-stimulating factor 2 (CSF2 or GM-CSF) which are cytokines secreted by T_h1 cells. Upon activation, M1 macrophages secrete inflammatory cytokines like TNF-α, IL-1β, IL-12 along with reactive oxygen species to destruct pathogens and nitric oxide (NO) to induce vasodilatation and migration of immune cells. They serve as antigen-presenting cells (APCs) to further induce immune cells. On the other hand, TGF-β, PGE2, IL4, and IL-10 induce M2 polarization of macrophages. M2 phenotype roles in the remodeling of the tissues upon damage via phagocyting overexpressed ECM molecules instead of pathogens and promote angiogenesis for tissue regeneration. They also use PDL-1 ligand or secrete anti-inflammatory IL-10 cytokine to inhibit cytotoxic T cells. IL-10 is secreted by MSCs and M2 macrophages (Saldaña et al. 2019), also activates anti-inflammatory regulatory T (T_{reg}) cells (Van Dalen et al. 2019). MSCs can direct monocytes and mature dendritic cells to an immature state, making them more susceptible to NK cells for degradation (Uccelli, Moretta, and Pistoia 2008).

B cells differentiate into antibody-producing plasma cells upon activation via APCs and IL-4 signaling. MHCs suppresses B cell proliferation and terminal differentiation (Asari et al. 2009) into plasma cells via controlling their both proliferation and differentiation (Corcione et al. 2006) by realizing the soluble factors such as TGFβ1 (Nicola et al. 2002), PGE2 (Aggarwal and Pittenger 2005), and IDO (Meisel et al. 2004). Still, with an unknown paracrine secretion, MSC can downregulate the Blimp-1 expression in B-cells and suppresses the B-cell terminal differentiation (Asari et al. 2009). Another mechanism to inhibit the B-cell proliferation is blocking the cell cycle at G_0/G_1 phase via the soluble inhibitory factors (Corcione et al. 2006). This also emphasizes the importance of the paracrine effects of MSCs.

T_{reg} cells are the anti-inflammatory mediators of the immune response working as an antagonist to T_h cells. MSCs execute their immunomodulatory effects (Vieira Paladino et al. 2017) by activating T_{reg} cells which in turn suppresses the immune response of cytotoxic T cells, NK cells, and macrophages. In other words, the indirect mechanism of inhibiting T cell function of MSCs is by stimulating the expansion of T_{reg} cells (D. Wang and Sun 2018). When T_{reg} and MSCs were cocultured, it was observed an increasing rate in the proliferation of T_{reg} cells via TGF-β secretion (D. Wang, Feng, et al. 2014). Additionally, MSCs regulate this indirect mechanism by adhering to T_h-17 cells and blocking their differentiation and inducing the T_{reg} cells (Ghannam et al. 2010).

3.2.2. Immunomodulatory Factors of MSCs

A wide range of immunomodulatory molecules is secreted by MSCs including IL-6, IL-8, IL-10, IL-12, IDO, IFN-γ, PGE2, VEGF, M-CSF, HGF and TGF-β1 (Naji et al. 2019; Karantalis and Hare 2015; Maumus, Jorgensen, and Noël 2013; Abumaree et al. 2012). IFN-γ is secreted by immune cells upon contacting with virus-infected cells which secretes IFN-α and IFN-β. IFN-γ is a mediator for the proliferation decrease of IL2, IL7 or IL15 activated lymphocytes. IFN-γ also affects the character of the MSCs to as immune-stimulant or suppressant phenotype, in a dose-dependent manner. However, the amount of IFN-γ, sufficient for character change is controversial and yet to be determined (Abumaree et al. 2012).

HLA-G, which is an MHC-I type receptor, expressed on the surface of the MSCs in membrane-bound form or secreted by them as soluble (HLA-G5) form, following IL-10 stimulation. NK cells, dendritic cells and CD8+ cytotoxic T cells (T_C cells) can interact and be suppressed by both membrane-bound and soluble isoforms of HLA-G (Abumaree et al. 2012).

PGE2 also blocks the differentiation of dendritic APCs (Spaggiari et al. 2009). Since macrophage activation syndrome (MAS) is one of the main causes of COVID-19 deaths, inhibition of monocyte-to-macrophage differentiation by MSCs can prevent MAS and calm the cytokine storm. M-CSF stimulated the proliferation of monocytes, therefore, block their macrophage transformation (Abumaree et al. 2012).

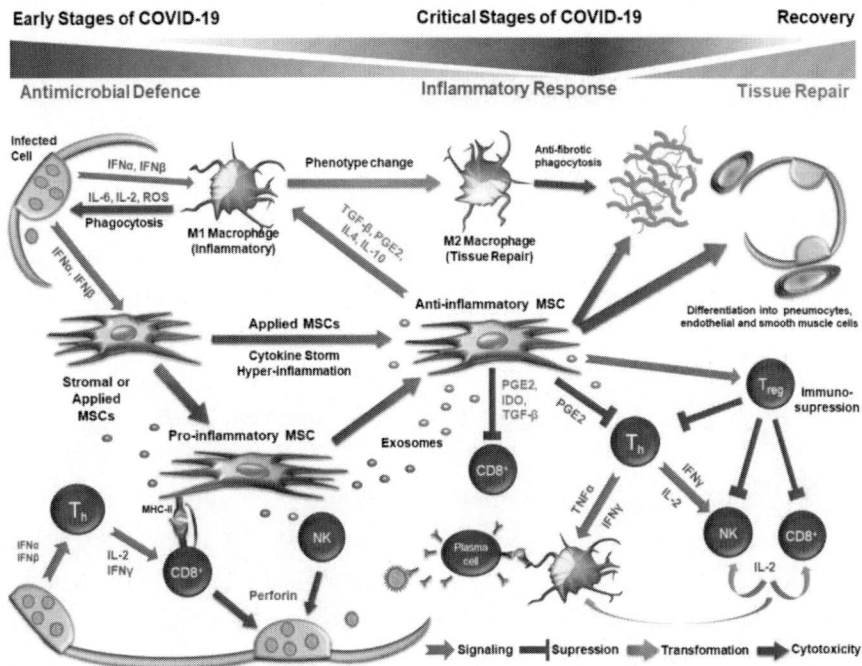

Figure 4. Immunomodulatory interactions of MSCs during COVID-19 pathogenesis. Stromal or transferred MSCs can be stimulated via INFα and IFNβ to turn into pro-inflammatory MSCs, which express MHC-II molecules and activate immune cells via mechanisms like receptor-receptor binding, cytokines or exosomes. However, during cytokine storm, resident or transplanted MSCs act as immunosuppressors via inhibiting Th or cytotoxic cells, activating Treg cells, or turning the macrophages to M2 phenotype. Anti-inflammatory MSCs also help the rebuilding process of damaged tissues.

Another important actor of the cytokine storm is IL-6, which is synthesized by macrophages, upon response to viral detection by IFN-γ, through the Notch-1 pathway. IL-6 can also be activated through positive feedback of itself (Rizzo et al. 2020). Our ongoing clinical trials also showed an increased number of IL-6 in patients, along with other pro-inflammatory cytokines, in preliminary results (unpublished data), which is also similar to the findings of others (Lei et al. 2020). IL-6 functions both as a pro- and anti-inflammatory molecule. Anti-inflammatory effects are mediated by soluble forms of IL-6 through the classical signaling mechanism, where IL-6 binds to the membrane receptor. However, pro-inflammatory effects emerge upon trans-signaling where IL-6 binds to sIL-6R molecule and attaches gp130 receptor. A monoclonal antibody drug, Tocilizumab (Roche), blocks both classical and trans pathways, therefore, inhibits both harmful and beneficial effects (C. Zhang et al. 2020) which was already passing through clinical trials for COVID-19 treatment (Clinical trial ID: ChiCTR2000029765) (Ye, Wang, and Mao 2020).

All these data point that MSCs have an overall inhibitory effect on the immune system, but this immune-suppressant effect is under the influence of other cytokines such as IFN-γ, and amount of these cytokines is also important for MSC actions which is summarized in Figure 4.

4. Previous and Current Clinical Uses of MSCs on COVID-19 Related Symptoms

Due to the various regenerative, anti-inflammatory, and properties of MSCs, they have been used in numerous clinical trials. There are currently more than 9000 entries in www.clinicaltrials.gov for MSC which ranges from acute injuries like burns and spinal cord cuts to chronic inflammatory or autoimmune diseases. Trials also include infections, cognitive problems, metabolic diseases, and cancer. All these studies exploit one of the mentioned properties of MSCs. We also demonstrated the benefits of MSC treatments with several clinical trials on various diseases like hypoxic-ischemic encephalopathy (Kabataş et al. 2018), cerebral palsy (Okur et al. 2018), Duchenne Muscular Distrophy (Dai et al. 2018).

COVID-19 causes symptoms like epithelial and alveolar damage in lungs, ARDS, hyper-inflammation, MAS, endothelial damage, DIC, and multiple organ damage. Extrapulmonary damaged organs including but not limited to the heart, brain, kidney, and intestines. Consequently, respiratory, cardiovascular, nervous, and digestive systems may need repair during or even after the disease. Here, we will focus on the MSC studies with COVID-19 related symptoms. Since there is an immense number of MSC clinical trials, we analyzed a small percent of them which we thought most related to COVID-19, like ARDS, pneumonia, infections, autoimmune disorders, and finally COVID-19 itself which are demonstrated on Table 1.

Table 1. All clinical trials, registered to the US National Library of Medicine Clinical Trials database related with directly COVID-19 itself or related conditions. Trials were listed as total number or with MSC only for each condition

Disease / Condition	Total Clinical Trials	Trials with MSCs	MSC Percentage
MSC	9447		
COVID-19	3237	238	7,4
ARDS	1314	80	6,1
Pneumonia	2350	109	4,6
Autoimmune	9387	265	2,8
Inflammation	6059	235	3,9
Respiratory	6947	246	3,5
Sepsis	2206	84	3,8
GVHD	874	47	5,4
Metabolic Disorders	22731	566	2,5

Source: www.clinicaltrials.gov Access date: 6 September 2020.

4.1. Pneumonia, Cystic Fibrosis and ARDS

Acute respiratory distress syndrome (ARDS) causes by the disruption of the blood-air barrier, starting with pulmonary edema during inflammation and have symptoms like hypoxemia, tachypnea, pulmonary infiltrates like the ones observed in the ground-glass images, and in the end stage, loss of lung function and death (Masterson et al. 2015). Despite decades of research, few therapeutic strategies for treating clinical ARDS have emerged (L. Liu et al. 2015). Regenerative properties of MSCs are attributed to their secretomes and exosomes. Many animal model studies developed during the last twenty years as they were summarized with detail in reviews (Worthington and Hagood 2020; Lopes-Pacheco et al.

2020). Almost all these animal models showed a decrease in lung inflammation, fibrosis, and pulmonary arterial hypertension, an increase in T_{reg} cell activity and character shift of macrophages to anti-inflammatory M2 phenotype, and increased survival of the animals. In an animal study of ours, performed in 2012, we observed increased lifespan of rats when MSCs applied in an experimental ARDS. We also detected decrease in proinflammatory cytokines, ILβ1, IL-6, MPO and MIP-2 and an increase in EP3 and IL1RA, a receptor for PGE signaling and antagonist of inflammatory IL-1A respectively, indicating anti-inflammatory effects of MSCs in ARDS (Yilmaz et al. 2013). In a recently published open-label, single-armed and early phase (1/2) study of ours also showed significant alleviations of the symptoms of chronic obstructive pulmonary disease (COPD) in a patient group with a mean age of 56 (E. Karaoz, Kalemci, and Ece 2020).

Another disease affecting lungs are cystic fibrosis (CF), which is an autosomal recessive disease resulting from mutations in Cystic Fibrosis Transmembrane Conductance Regulator (CFTR) gene. This defect leads to mucous obstruction, infection, inflammation, and progressive lung damage (Caretti et al. 2019), which are similar to COVID-19 symptoms. In vitro and animal models (Zulueta et al. 2018; Sutton et al. 2017) showed the effectiveness of the MSC in CF and it is also being tested in a clinical trial (NCT02866721). Relation between the effects of MSCs and symptoms of CF requires deeper investigation of the subject with more clinical trials.

4.2. Autoimmune and Chronic Inflammatory Diseases

Autoimmune disorders are a wide range group that consist of over 80 different diseases, that are occurred since the immune system cannot distinguish between its self and non-self (L. Wang, Wang, and Gershwin 2015). In other words, they are antibody-mediated diseases, where immune cells lock on certain cells via their antigen-antibody binding and start inflammatory cytokine secretions. Various organs or tissues may be affected by the immunological attacks therefore symptoms differ from each other (L. Wang, Wang, and Gershwin 2015). Due to the mentioned immunosuppression characteristics of the MSCs, they act as are novel therapeutics that regulate immune cells against autoimmune diseases (Tyndall 2011; Castro-Manrreza and Montesinos 2015) where antibody treatments remain insufficient (Pérez-De-Lis et al. 2017). As mentioned in Figure 5, 265 clinical trials were registered as MSC treatment for autoimmune diseases in clinicaltrials.gov. Their treatment effect to autoimmune diseases is thought to be through their paracrine secretions like cytokines and exosomes (Fattore et al. 2015; Fontaine et al. 2016; Abbasi-Malati et al. 2018) rather than cell-cell interaction. Within the MSCs found promising for clinical applications of some autoimmune diseases such as Lupus Erythematous (D. Wang, Li, et al. 2014), Sjögren syndrome (Shi et al. 2018), Crohn's Disease (Carvello et al. 2019; Freedman et al. 2010) and MS (Dulamea 2015). The main mechanism behind the MSC therapy in whole autoimmune diseases is regulating the maturation lineage of NK, T/B cells and dendritic cells via releasing miRNA containing exosomes (Blanco et al. 2016; Fontaine et al. 2016). These exosomes provide immune synapsis between antigen presenting cells and T-cells, alter cellular functions and correct the misguided targets (Quesenberry et al. 2015; Fontaine et al. 2016).

GvHD is an autoimmune disorder where a complicated severe immune reaction can emerge after allogeneic hematopoietic stem cell transplantation. Suppression of T-cell proliferation via apoptotic genes, released immunosuppressive cytokines and proliferation of Il-17A-producing Tregs are the key mediators in both the prevention and the treatment of GvHD. In addition to our molecular level assessment, Tarifa et al. published a detailed systemic review of MSC transplants to GvHD patients between 2004-2019 (Morata-Tarifa et al. 2020). A total 654 GvHD patients were evaluated and increased survival rates, decreased incidence and organ responses were indicated that allogeneic- MSC could be powerful therapeutic to both prevent and treat GvHD. MSC transplantations have begun to be routinely used for prevention of GvHD following the allogeneic transplantation. As it also mentioned in Table 1, 47 clinical trials were registered to in clinicaltrials.gov for GvHD using MSCs.

4.3. Regenerative Abilities

Various studies were conducted to determine whether stem cell-based treatments can be used to regenerate damaged tissues like endothelial or alveolar cells which are the targets of SARS-CoV-2 long before the pandemic. One of them explored the potential for differentiation of MSCs into endothelial cells by recombinantly produced VEGF-A, which plays an important role in cell differentiation and proliferation. The study conducted suggested that mesenchymal stem cell-based treatments can be used to support endothelial regeneration (Khaki et al. 2018).

A 2019 study showed that MSCs overexpressing p130/E2F4 genes helped the repair of damaged lung tissue by differentiating into ATII cells in LPS-induced ARDS mice (X. Zhang et al. 2019). A study with human cells on decellularized rodent lungs demonstrated differentiation of adipose tissue derived hMSCs into ATII and Clara cell-like cells (Mendez et al. 2014). An organoid study also showed that MSCs enhance differentiation of lung epithelial stem cells into alveolar cells in 3D cultures, indicating factors secreted from MSCs may replace missing signals and activate epithelial cell repair (Leeman et al. 2019).

4.4. Clinical Applications of MSCs on SARS, MERS and COVID-19 Cases

Since SARS and MERS affected a relatively low number of people, we did not encounter any MSC applications for that outbreaks. However, there are several in vitro and animal studies for avian influenza virus strains H1N1 (Darwish et al. 2013; Chan et al. 2016), H5N1 (Chan et al. 2016), H9N2 (Y. Li et al. 2016). MSCs can improve lung function through anti-inflammatory effects in acute injury lung in a mouse models (B. Wang et al. 2017; Xiang et al. 2017; Hu and Li 2018). In a human trial, allogeneic MSC treatment was tested in 9 patients with ARDS, and there were no prespecified adverse events including hypoxemia, cardiac arrhythmia, and ventricular tachycardia was reported (Horie et al. 2018). Current trials registered in clinicaltrials.gov (search date 5 September 2020) are shown in Table 2. This is quite gratifying because it shows that the stem cell treatment alternative in COVID-19 has been progressing to further stages.

Table 2. The recruitment status of all MSC treatments in COVID-19 clinical trials with their development phases. A total of 152 registered studies were found to use MSCs for COVID-19 while only one of them, a phase II study, were completed (www.clinicaltrials.gov Access date: 6 September 2020)

Clinical Trial Phase	Number
All Phase 1 Studies	48
MSC | Not yet recruiting | COVID-19 | Phase 1	21
MSC | Recruiting | COVID-19 | Phase 1	22
MSC | Active | COVID-19 | Phase 1	3
MSC | Enrolling by invitation | COVID-19 | Phase 1	2
All Phase 2 Studies	69
MSC | Enrolling by invitation | COVID-19 | Phase 2	2
MSC | Not yet recruiting | COVID-19 | Phase 2	22
MSC | Recruiting | COVID-19 | Phase 2	43
MSC | Active, not recruiting | COVID-19 | Phase 2	1
MSC | Completed | COVID-19 | Phase 2	1
All Phase 3 Studies	35
MSC | Not yet recruiting | COVID-19 | Phase 3	12
MSC | Recruiting | COVID-19 | Phase 3	22
MSC | Active, not recruiting | COVID-19 | Phase 3	1
TOTAL	152

Recently, Leng et al. reported the pilot trial of intravenous MSC transplantation. It was performed on seven patients with COVID-19 infected pneumonia, including one critically severe type, four severe types, and two common types, also, the study had three patients as placebo controls. Reportedly, the symptoms of high fever (38.5°C ± 0.5°C), weakness, shortness of breath, and low oxygen saturation disappeared 2~4 days after the transplantation of MSCs. Meanwhile, right after the infusion of MSCs, there was no acute infusion-related or allergic reactions in two hours after transplantation. The authors claimed that the MSC treatment significantly improved the inflammation in severe COVID-19 patients and they attributed this effect to the immunosuppression capacity of this cell-based treatment, as demonstrated by reduced levels of proinflammatory cytokines and chemokines and increased IL-10 and VEGF contributed to the lung repair. They also concluded that MSCs improved the outcome of COVID-2019 patients may be in dual effect: regulating inflammatory response and promoting tissue repair and regeneration (Leng et al. 2020). This study has a quite limited number of patients and very short term follow up protocol. There are earlier reports from 2014 (Zheng et al. 2014) and 2015 (Wilson et al. 2015) of two clinical trials testing the safety of MSC infusion in ARDS patients demonstrated no adverse effects of cell therapy and some limited treatment efficacy. The ongoing clinical trials, especially the ones having randomized controlled design, which is demonstrated in Table 2, are expected to have published reports soon. We believe that these results will help the scientific community to have the evidence to prove the MSC treatment's beneficial outcome for COVID-19 cases.

4.5. Our Preliminary Results of MSC Therapy for COVID-19 Patients

During the preparation of this paper, our clinical trials (ID: NCT04392778) were close to the conclusion. We applied MSC treatment with permission from Turkey's Ministry of Health for each patient. All patients were intubated and in a critical stage. MSCs were applied as 3 million cells/kg via IV route at with three days interval for three times. Our preliminary results showed that out of two-thirds of the patients in pre-ex condition were saved upon stem cell treatment. One of these patients was under the effect of cytokine storm with high ferritin and d-dimer levels and showed signs of multiple organ damage including occlusions in the brain stem. We applied three doses of MSCs via IV route and one additional intrathecal application upon detection of brain involvement. The patient was eventually saved and discharged from the hospital.

5. Timing for Stem Cell Treatment

Since around 80% of COVID-19 patients experience zero or mild effects (Day 2020), they do not require stem cell treatment either. Therapy requirements start with the hospitalization of the patient. During the first, viral replication phase of the disease, antiviral drugs seem useful. Since the immune system actively fights the virus during the first phase, immunesuppressant drugs like IL-6 inhibitors should be avoided. Therefore, MSC treatment may seem unnecessary at this point due to their immunosuppressant effects. However, MSCs may also act as pro-inflammatory agents, depending on the IFNγ levels (Abumaree et al. 2012). This effect of MSCs is broadly tested in clinical trials, usually for cancer treatments. Additionally, MSCs have antimicrobial effects through their secreted molecules like hepcidin, which is showed to be effective against Hepatitis C Virus via altering intracellular antiviral response (H. Liu et al. 2012); β-defensins, which target viral envelope and capsid proteins; LL-37, which is shown to be effective against respiratory syncytial virus (Currie et al. 2013) and also seasonal and pandemic influenza A viruses (Tripathi et al. 2015). Therefore, MSCs can be an effective and natural treatment alternative directly against SARS-CoV-2. During the proposed second phase of the disease, where the viral replication ceases but cytokine storm is taking over (Siddiqi and Mehra 2020), MSCs can be used exploiting their anti-inflammatory effects, which is discussed above.

MSC treatment for organs other than lungs has a major drawback called 'the first pass effect', described as the entanglement of the MSCs in umbilical capillaries. Since the lung capillaries are the first narrow passages for cells to pass upon venous administration of MSCs, the majority traps within lungs (Fischer et al. 2009). Although this may be a huge disadvantage for other organ treatments, in COVID-19 cases where the lungs are the most affected organs, these obstacle turns into an advantage for the homing of the MSCs into the most injured area.

Although MSCs are effective agents against various symptoms in numerous diseases, they carry out these functions through exosomes (Qiu et al. 2018). Exosomes contain molecules like signaling peptides, and miRNA to perform signal transduction. However, MSC reaction and response to the inflammation are dependent on the environmental stimulants and therefore, secreted exosomes ingredients are affected by the environment (Yin, Wang, and Zhao 2019). Although there are five ongoing trials with exosomes against COVID-19 and many more for other diseases, to fully benefit their therapeutic potential the exosome characteristics must be adjusted for the disease-specific environment. To accomplish this goal, stem cells may be held within a similar environment or the targeted situation like high cytokine levels of the body to mimic MAS and force MSCs to secrete exosomes carrying anti-inflammatory molecules.

Consequently, MSCs may be used during the first phase of hospitalization of COVID-19 patients due to their antimicrobial and pro-inflammatory effects. The same cells can also act as anti-inflammatory agents and can be effectively used to calm down the cytokine storm. There is also a possibility that the MSCs, transferred during the viral replication phase for antimicrobial effects, may remain within the tissues of the patient to ameliorate or even fully prevent cytokine storm.

However, we suggest, physicians should closely follow the patients and determine the therapy needs according to the developing symptoms and lab results. The number of stem cells during infusion should also be determined carefully by specialists since high doses of MSCs can accumulate in lung capillaries and may cause pulmonary embolism (Y. Y. Liu et al. 2017) which may endanger the patients who are already under risk of coagulopathy. Stem cell treatment can also be combined with anticoagulants like heparin which is already under use for COVID-19 treatments (Liao et al. 2017; Tang, Bai, et al. 2020).

CONCLUSION AND FUTURE DIRECTIONS

COVID-19 marked 2020 with its global effects on every aspect of life. Among these, it caused deaths of more than 900.000 people within a few months. As discussed above, SARS-CoV-2 not only affects the lungs but also harm endothelial cells of the other organs. Following the viral infection, hyper-inflammation and cytokine storm causes additional and probably deadlier effects. Therefore, causes of these deaths were differing as ARDS, multiple organ failure, MI, and strokes due to inflammation, MAS, and coagulopathy (Zaim et al. 2020).

On the other hand, MSCs were used numerous clinical trials due to their regenerative, anti-inflammatory, and immunosuppressant properties. As we pointed on the cellular and molecular level with various data, these effects may benefit from damage caused by COVID-19. Regenerative abilities can be used to replenish damaged tissues of lungs and other organs while anti-inflammatory properties reduce lung damage and immune-suppressant properties calm down cytokine storm. These aspects were proven by previous clinical trials showed that MSCs were already effective on these symptoms. Current clinical trial outcomes for COVID-19 also support this hypothesis. Therefore, we propose that MSC therapies can be used to treat COVID-19 patients effectively along with other pharmaceutical treatment approaches, considering the stage of the diseases and organs under the damaging effects of the disease. As we have reviewed in a recent report (Irmak, Darıcı, and Karaöz 2020), the safety and efficacy in the long term remain to be proven via more clinical trials with higher number of enrolled patients to confirm the results. If the severity of the COVID-19 continues with second or third waves, cGMP certificated MSC production and treatment centers can be established for mass treatment of current and future pandemics.

ACKNOWLEDGMENTS

We would like to thank Öznur BAYGUT for her help in this chapter.

REFERENCES

Abbasi-Malati, Zahra, Amaneh Mohammadi Roushandeh, Yoshikazu Kuwahara, and Mehryar Habibi Roudkenar. 2018. "Mesenchymal

Stem Cells on Horizon: A New Arsenal of Therapeutic Agents." *Stem Cell Reviews and Reports*. Humana Press Inc. https://doi.org/10.1007/s12015-018-9817-x.

Abumaree, Mohamed, Mohammed Al Jumah, Rishika A. Pace, and Bill Kalionis. 2012. "Immunosuppressive Properties of Mesenchymal Stem Cells." *Stem Cell Reviews and Reports* 8 (2): 375–92. https://doi.org/10.1007/s12015-011-9312-0.

Aggarwal, Sudeepta, and Mark F. Pittenger. 2005. "Human Mesenchymal Stem Cells Modulate Allogeneic Immune Cell Responses." *Blood* 105 (4): 1815–22. https://doi.org/10.1182/blood-2004-04-1559.

Ai, Tao, Zhenlu Yang, Hongyan Hou, Chenao Zhan, Chong Chen, Wenzhi Lv, Qian Tao, Ziyong Sun, and Liming Xia. 2020. "Correlation of Chest CT and RT-PCR Testing in Coronavirus Disease 2019 (COVID-19) in China: A Report of 1014 Cases." *Radiology* 2019: 200642. https://doi.org/10.1148/radiol.2020200642.

Akdis, Mübeccel, Alar Aab, Can Altunbulakli, Kursat Azkur, Rita A. Costa, Reto Crameri, Su Duan, et al. 2016. "Interleukins (from IL-1 to IL-38), Interferons, Transforming Growth Factor β, and TNF-α: Receptors, Functions, and Roles in Diseases." *Journal of Allergy and Clinical Immunology* 138 (4): 984–1010. https://doi.org/10.1016/j.jaci.2016.06.033.

Akker, F. van den, K. R. Vrijsen, J. C. Deddens, J. W. Buikema, M. Mokry, L. W. van Laake, P. A. Doevendans, and J. P.G. Sluijter. 2018. "Suppression of T Cells by Mesenchymal and Cardiac Progenitor Cells Is Partly Mediated via Extracellular Vesicles." *Heliyon* 4 (6). https://doi.org/10.1016/j.heliyon.2018.e00642.

Asari, Sadaki, Shin Itakura, Kevin Ferreri, Chih Pin Liu, Yoshikazu Kuroda, Fouad Kandeel, and Yoko Mullen. 2009. "Mesenchymal Stem Cells Suppress B-Cell Terminal Differentiation." *Experimental Hematology* 37 (5): 604–15. https://doi.org/10.1016/j.exphem.2009.01.005.

Barnes, Betsy J., Jose M. Adrover, Amelia Baxter-Stoltzfus, Alain Borczuk, Jonathan Cools-Lartigue, James M. Crawford, Juliane Daßler-Plenker, et al. 2020. "Targeting Potential Drivers of COVID-

19: Neutrophil Extracellular Traps." *Journal of Experimental Medicine* 217 (6): 1–7. https://doi.org/10.1084/jem.20200652.

Blanc, K. Le. 2006. "Mesenchymal Stromal Cells: Tissue Repair and Immune Modulation." *Cytotherapy* 8 (6): 559–61. https://doi.org/10.1080/14653240601045399.

Blanco, Belén, María del Carmen Herrero-Sánchez, Concepción Rodríguez-Serrano, María Lourdes García-Martínez, Juan F. Blanco, Sandra Muntión, Mariano García-Arranz, Fermín Sánchez-Guijo, and Consuelo del Cañizo. 2016. "Immunomodulatory Effects of Bone Marrow versus Adipose Tissue-Derived Mesenchymal Stromal Cells on NK Cells: Implications in the Transplantation Setting." *European Journal of Haematology* 97 (6): 528–37. https://doi.org/10.1111/ejh.12765.

Bruce Aylward (WHO); Wannian Liang (PRC). 2020. "Report of the WHO-China Joint Mission on Coronavirus Disease 2019 (COVID-19)." *The WHO-China Joint Mission on Coronavirus Disease 2019*. Vol. 1.

Caplan, Arnold I. 2007. "Adult Mesenchymal Stem Cells for Tissue Engineering versus Regenerative Medicine." *Journal of Cellular Physiology* 213 (2): 341–47. https://doi.org/10.1002/jcp.21200.

———. 2017. "Mesenchymal Stem Cells: Time to Change the Name!" *Stem Cells Translational Medicine* 6 (6): 1445–51. https://doi.org/10.1002/sctm.17-0051.

Caretti, Anna, Valeria Peli, Michela Colombo, and Aida Zulueta. 2019. "Lights and Shadows in the Use of Mesenchymal Stem Cells in Lung Inflammation, a Poorly Investigated Topic in Cystic Fibrosis." *Cells* 9 (1): 20. https://doi.org/10.3390/cells9010020.

Carvello, Michele, Amy Lightner, Takayuki Yamamoto, Paulo Gustavo Kotze, and Antonino Spinelli. 2019. "Mesenchymal Stem Cells for Perianal Crohn's Disease." *Cells* 8 (7): 764. https://doi.org/10.3390/cells8070764.

Castro-Manrreza, Marta E., and Juan J. Montesinos. 2015. "Immunoregulation by Mesenchymal Stem Cells: Biological Aspects and Clinical Applications." *Journal of Immunology Research.* https://doi.org/10.1155/2015/394917.

CDC. 2020. *Symptoms of Coronavirus.* CDC. 2020.

Chan, Michael C.W., Denise I.T. Kuok, Connie Y.H. Leung, Kenrie P.Y. Hui, Sophie A. Valkenburg, Eric H.Y. Lau, John M. Nicholls, et al. 2016. "Human Mesenchymal Stromal Cells Reduce Influenza A H5N1-Associated Acute Lung Injury in Vitro and in Vivo." *Proceedings of the National Academy of Sciences of the United States of America* 113 (13): 3621–26. https://doi.org/10.1073/pnas.1601911113.

Corcione, Anna, Federica Benvenuto, Elisa Ferretti, Debora Giunti, Valentina Cappiello, Francesco Cazzanti, Marco Risso, et al. 2006. "Human Mesenchymal Stem Cells Modulate B-Cell Functions." *Blood* 107 (1): 367–72. https://doi.org/10.1182/blood-2005-07-2657.

Currie, Silke M., Emily Gwyer Findlay, Brian J. McHugh, Annie Mackellar, Tian Man, Derek Macmillan, Hongwei Wang, Paul M. Fitch, Jürgen Schwarze, and Donald J. Davidson. 2013. "The Human Cathelicidin LL-37 Has Antiviral Activity against Respiratory Syncytial Virus." *PLoS ONE* 8 (8). https://doi.org/10.1371/journal.pone.0073659.

Dai, Alper, Osman Baspinar, Ahmet Yeşilyurt, Eda Sun, Çiğdem İnci Aydemir, Olga Nehir Öztel, Davut Unsal Capkan, Ferda Pinarli, Abdullah Agar, and Erdal Karaöz. 2018. "Efficacy of Stem Cell Therapy in Ambulatory and Nonambulatory Children with Duchenne Muscular Dystrophy - Phase I-II." *Degenerative Neurological and Neuromuscular Disease* 8: 63–77. https://doi.org/10.2147/DNND.S170087.

Dalen, Floris J. Van, Marleen H.M.E. Van Stevendaal, Felix L. Fennemann, Martijn Verdoes, and Olga Ilina. 2019. "Molecular Repolarisation of Tumour-Associated Macrophages." *Molecules* 24 (1). https://doi.org/10.3390/molecules24010009.

Darwish, Ilyse, David Banner, Samira Mubareka, Hani Kim, Rickvinder Besla, David J. Kelvin, Kevin C. Kain, and W. Conrad Liles. 2013. "Mesenchymal Stromal (Stem) Cell Therapy Fails to Improve Outcomes in Experimental Severe Influenza." *PLoS ONE* 8 (8). https://doi.org/10.1371/journal.pone.0071761.

Dawood, Fatimah S., A. Danielle Iuliano, Carrie Reed, Martin I. Meltzer, David K. Shay, Po Yung Cheng, Don Bandaranayake, et al. 2012. "Estimated Global Mortality Associated with the First 12 Months of 2009 Pandemic Influenza A H1N1 Virus Circulation: A Modelling Study." *The Lancet Infectious Diseases* 12 (9): 687–95. https://doi.org/10.1016/S1473-3099(12)70121-4.

Day, Michael. 2020. "Covid-19: Identifying and Isolating Asymptomatic People Helped Eliminate Virus in Italian Village." *BMJ (Clinical Research Ed.)* 368: m1165. https://doi.org/10.1136/bmj.m1165.

Demircan, Pinar Cetinalp, Ayla Eker Sariboyaci, Zehra Seda Unal, Gulcin Gacar, Cansu Subasi, and Erdal Karaoz. 2011. "Immunoregulatory Effects of Human Dental Pulp-Derived Stem Cells on T Cells: Comparison of Transwell Co-Culture and Mixed Lymphocyte Reaction Systems." *Cytotherapy* 13 (10): 1205–20. https://doi.org/10.3109/14653249.2011.605351.

Dong, Ensheng, Hongru Du, and Lauren Gardner. 2020. "An Interactive Web-Based Dashboard to Track COVID-19 in Real Time." *The Lancet Infectious Diseases* 20 (5): 533–34. https://doi.org/10.1016/S1473-3099(20)30120-1.

Dotiwala, Farokh, Sachin Mulik, Rafael B. Polidoro, James A. Ansara, Barbara A. Burleigh, Michael Walch, Ricardo T. Gazzinelli, and Judy Lieberman. 2016. "Killer Lymphocytes Use Granulysin, Perforin and Granzymes to Kill Intracellular Parasites." *Nature Medicine* 22 (2): 210–16. https://doi.org/10.1038/nm.4023.

Dulamea, A. 2015. "Mesenchymal Stem Cells in Multiple Sclerosis - Translation to Clinical Trials." *Journal of Medicine and Life*. Carol Davila - University Press.

English, Coralie, and Susan L Hillier. 2010. "Circuit Class Therapy for Improving Mobility after Stroke." In *Cochrane Database of Systematic*

Reviews. Vol. 2010. John Wiley & Sons, Ltd. https://doi.org/10.1002/14651858.cd007513.pub2.

Fanelli, Vito, Aikaterini Vlachou, Shirin Ghannadian, Umberto Simonetti, Arthur S. Slutsky, and Haibo Zhang. 2013. "Acute Respiratory Distress Syndrome: New Definition, Current and Future Therapeutic Options." *Journal of Thoracic Disease*. J Thorac Dis. https://doi.org/10.3978/j.issn.2072-1439.2013.04.05.

Fattore, Andrea Del, Rosa Luciano, Luisa Pascucci, Bianca Maria Goffredo, Ezio Giorda, Margherita Scapaticci, Alessandra Fierabracci, and Maurizio Muraca. 2015. "Immunoregulatory Effects of Mesenchymal Stem Cell-Derived Extracellular Vesicles on T Lymphocytes." *Cell Transplantation* 24 (12): 2615–27. https://doi.org/10.3727/096368915X687543.

Fischer, Uwe M, Matthew T Harting, Fernando Jimenez, Werner O Monzon-Posadas, Hasen Xue, Sean I Savitz, Glen A Laine, and Charles S Cox. 2009. "Pulmonary Passage Is a Major Obstacle for Intravenous Stem Cell Delivery: The Pulmonary First-Pass Effect." *Stem Cells and Development* 18 (5): 683–92. https://doi.org/10.1089/scd.2008.0253.

Fodoulian, Leon, Joel Tuberosa, Daniel Rossier, Basile Landis, Alan Carleton, and Ivan Rodriguez. 2020. "SARS-CoV-2 Receptor and Entry Genes Are Expressed by Sustentacular Cells in the Human Olfactory Neuroepithelium." *BioRxiv*, no. April: 2020.03.31.013268. https://doi.org/10.1101/2020.03.31.013268.

Fontaine, Magali J., Hank Shih, Richard Schäfer, and Mark F. Pittenger. 2016. "Unraveling the Mesenchymal Stromal Cells' Paracrine Immunomodulatory Effects." *Transfusion Medicine Reviews*. W.B. Saunders. https://doi.org/10.1016/j.tmrv.2015.11.004.

Freedman, Mark S., Amit Bar-Or, Harold L. Atkins, Dimitrios Karussis, Francesco Frassoni, Hillard Lazarus, Neil Scolding, Shimon Slavin, Katarina Le Blanc, and Antonio Uccelli. 2010. "The Therapeutic Potential of Mesenchymal Stem Cell Transplantation as a Treatment for Multiple Sclerosis: Consensus Report of the International MSCT

Study Group." In *Multiple Sclerosis*, 16:503–10. Mult Scler. https://doi.org/10.1177/1352458509359727.

Gafa, Valérie, Roberto Lande, Maria Cristina Gagliardi, Martina Severa, Elena Giacomini, Maria Elena Remoli, Roberto Nisini, et al. 2006. "Human Dendritic Cells Following Aspergillus Fumigatus Infection Express the CCR7 Receptor and a Differential Pattern of Interleukin-12 (IL-12), IL-23, and IL-27 Cytokines, Which Lead to a Th1 Response." *Infection and Immunity* 74 (3): 1480–89. https://doi.org/10.1128/IAI.74.3.1480-1489.2006.

Ge, Xing Yi, Jia Lu Li, Xing Lou Yang, Aleksei A. Chmura, Guangjian Zhu, Jonathan H. Epstein, Jonna K. Mazet, et al. 2013. "Isolation and Characterization of a Bat SARS-like Coronavirus That Uses the ACE2 Receptor." *Nature* 503 (7477): 535–38. https://doi.org/10.1038/nature 12711.

Ghannam, Soufiane, Jérôme Pène, Gabriel Torcy-Moquet, Christian Jorgensen, and Hans Yssel. 2010. "Mesenchymal Stem Cells Inhibit Human Th17 Cell Differentiation and Function and Induce a T Regulatory Cell Phenotype." *The Journal of Immunology* 185 (1): 302–12. https://doi.org/10.4049/jimmunol.0902007.

Gheblawi, Mahmoud, Kaiming Wang, Anissa Viveiros, Quynh Nguyen, Jiu Chang Zhong, Anthony J. Turner, Mohan K. Raizada, Maria B. Grant, and Gavin Y. Oudit. 2020. "Angiotensin-Converting Enzyme 2: SARS-CoV-2 Receptor and Regulator of the Renin-Angiotensin System: Celebrating the 20th Anniversary of the Discovery of ACE2." *Circulation Research*. NLM (Medline). https://doi.org/10.1161/ CIRCRESAHA.120.317015.

Gorbalenya, Alexander E., Susan C. Baker, Ralph S. Baric, Raoul J. de Groot, Christian Drosten, Anastasia A. Gulyaeva, Bart L. Haagmans, et al. 2020. "The Species Severe Acute Respiratory Syndrome-Related Coronavirus: Classifying 2019-NCoV and Naming It SARS-CoV-2." *Nature Microbiology* 5 (4): 536–44. https://doi.org/10.1038/s41564-020-0695-z.

Guo, Yan Rong, Qing Dong Cao, Zhong Si Hong, Yuan Yang Tan, Shou Deng Chen, Hong Jun Jin, Kai Sen Tan, De Yun Wang, and Yan Yan.

2020. "The Origin, Transmission and Clinical Therapies on Coronavirus Disease 2019 (COVID-19) Outbreak- A n Update on the Status." *Military Medical Research* 7 (1): 1–10. https://doi.org/10.1186/s40779-020-00240-0.

Habibzadeh, Parham, and Emily K. Stoneman. 2020. "The Novel Coronavirus: A Bird's Eye View." *International Journal of Occupational and Environmental Medicine* 11 (2): 65–71. https://doi.org/10.15171/ijoem.2020.1921.

Henderson, Lauren A., Scott W. Canna, Grant S. Schulert, Stefano Volpi, Pui Y. Lee, Kate F. Kernan, Roberto Caricchio, et al. 2020. "On the Alert for Cytokine Storm: Immunopathology in COVID-19." *Arthritis & Rheumatology*, 0–3. https://doi.org/10.1002/art.41285.

Hess, David C, Wael Eldahshan, and Elizabeth Rutkowski. 2020. *COVID-19-Related Stroke*, 1–4.

Hoffmann, Markus, Hannah Kleine-Weber, Simon Schroeder, Nadine Krüger, Tanja Herrler, Sandra Erichsen, Tobias S. Schiergens, et al. 2020. "SARS-CoV-2 Cell Entry Depends on ACE2 and TMPRSS2 and Is Blocked by a Clinically Proven Protease Inhibitor." *Cell* 181 (2): 271-280.e8. https://doi.org/10.1016/j.cell.2020.02.052.

Horie, Shahd, Hector Esteban Gonzalez, John G. Laffey, and Claire H. Masterson. 2018. "Cell Therapy in Acute Respiratory Distress Syndrome." *Journal of Thoracic Disease* 10 (9): 5607–20. https://doi.org/10.21037/jtd.2018.08.28.

Hu, Chenxia, and Lanjuan Li. 2018. "Preconditioning Influences Mesenchymal Stem Cell Properties in Vitro and in Vivo." *Journal of Cellular and Molecular Medicine*. Blackwell Publishing Inc. https://doi.org/10.1111/jcmm.13492.

Irmak, Duygu Koyuncu, Hakan Darıcı, and Erdal Karaöz. 2020. "Stem Cell Based Therapy Option in COVID-19: Is It Really Promising?" *Aging and Disease* 11 (4): 1–17. https://doi.org/http://dx.doi.org/10.14336/AD.2020.0608 Stem.

Jayaraman, Pukana, Prakash Nathan, Punitha Vasanthan, Sabri Musa, and Vijayendran Govindasamy. 2013. "Stem Cells Conditioned Medium:

A New Approach to Skin Wound Healing Management." *Cell Biology International* 37 (10): 1122–28. https://doi.org/10.1002/cbin.10138.

Jia, Hong Peng, Dwight C. Look, Lei Shi, Melissa Hickey, Lecia Pewe, Jason Netland, Michael Farzan, Christine Wohlford-Lenane, Stanley Perlman, and Paul B. McCray. 2005. "ACE2 Receptor Expression and Severe Acute Respiratory Syndrome Coronavirus Infection Depend on Differentiation of Human Airway Epithelia." *Journal of Virology* 79 (23): 14614–21. https://doi.org/10.1128/jvi.79.23.14614-14621.2005.

Kabatas, S., C. S. Demir, E. Civelek, I. Yilmaz, A. Kircelli, C. Yilmaz, Y. Akyuva, and E. Karaoz. 2018. "Neuronal Regeneration in Injured Rat Spinal Cord after Human Dental Pulp Derived Neural Crest Stem Cell Transplantation." *Bratislava Medical Journal* 119 (3): 143–51. https://doi.org/10.4149/BLL_2018_028.

Kabataş, Serdar, Erdinç Civelek, Çiğdem İnci, Ebru Yılmaz Yalçınkaya, Gülşen Günel, Gülay Kır, Esra Albayrak, Erek Öztürk, Gökhan Adaş, and Erdal Karaöz. 2018. "Wharton's Jelly-Derived Mesenchymal Stem Cell Transplantation in a Patient with Hypoxic-Ischemic Encephalopathy: A Pilot Study." *Cell Transplantation* 27 (10): 1425–33. https://doi.org/10.1177/0963689718786692.

Kampf, G., D. Todt, S. Pfaender, and E. Steinmann. 2020. "Persistence of Coronaviruses on Inanimate Surfaces and Their Inactivation with Biocidal Agents." *Journal of Hospital Infection*. W.B. Saunders Ltd. https://doi.org/10.1016/j.jhin.2020.01.022.

Karantalis, Vasileios, and Joshua M. Hare. 2015. "Use of Mesenchymal Stem Cells for Therapy of Cardiac Disease." *Circulation Research*. Lippincott Williams and Wilkins. https://doi.org/10.1161/CIRCRESAHA.116.303614.

Karaoz, E., S. Kalemci, and F. Ece. 2020. "Improving Effects of Mesenchymal Stem Cells on Symptoms of Chronic Obstructive Pulmonary Disease." *Bratislava Medical Journal* 121 (3): 188–91. https://doi.org/10.4149/BLL_2020_028.

Karaoz, Erdal, Pinar Cetinalp Demircan, Gulay Erman, Eda Gungorurler, and Ayla Eker Sariboyaci. 2017. "Comparative Analyses of Immunosuppressive Characteristics of Bone-Marrow, Wharton's Jelly, and Adipose Tissue-Derived Human Mesenchymal Stem Cells." *Turkish Journal of Haematology : Official Journal of Turkish Society of Haematology* 34: 213–25. https://doi.org/10.4274/tjh.2016.0171.

Karaöz, Erdal, Pinar Cetinalp Demircan, Özlem Saflam, Ayca Aksoy, Figen Kaymaz, and Gökhan Duruksu. 2011. "Human Dental Pulp Stem Cells Demonstrate Better Neural and Epithelial Stem Cell Properties than Bone Marrow-Derived Mesenchymal Stem Cells." *Histochemistry and Cell Biology* 136 (4): 455–73. https://doi.org/10.1007/s00418-011-0858-3.

Karaoz, Erdal, Filiz Tepekoy, Irem Yilmaz, Cansu Subasi, and Serdar Kabatas. 2019. "Reduction of Inflammation and Enhancement of Motility after Pancreatic Islet Derived Stem Cell Transplantation Following Spinal Cord Injury." *Journal of Korean Neurosurgical Society* 62 (2): 153–65. https://doi.org/10.3340/jkns.2018.0035.

Khaki, Mohsen, Ali Hatef Salmanian, Hamid Abtahi, Ali Ganji, and Ghasem Mosayebi. 2018. "Mesenchymal Stem Cells Differentiate to Endothelial Cells Using Recombinant Vascular Endothelial Growth Factor -A." *Reports of Biochemistry & Molecular Biology* 6 (2): 144–50. https://pubmed.ncbi.nlm.nih.gov/29761109.

Krishan Gupta, Sanjay Kumar Mohanty, Siddhant Kalra, Aayushi Mittal, Tripti Mishra, Jatin Ahuja, Debarka Sengupta, Gaurav Ahuja. 2020. *The Molecular Basis of Loss of Smell in 2019-NCoV Infected Individuals*, March. https://doi.org/10.21203/RS.3.RS-19884/V1.

Lai, Chih Cheng, Tzu Ping Shih, Wen Chien Ko, Hung Jen Tang, and Po Ren Hsueh. 2020. "Severe Acute Respiratory Syndrome Coronavirus 2 (SARS-CoV-2) and Coronavirus Disease-2019 (COVID-19): The Epidemic and the Challenges." *International Journal of Antimicrobial Agents* 55 (3): 105924. https://doi.org/10.1016/j.ijantimicag.2020.105924.

Lamers, Mart M., Joep Beumer, Jelte van der Vaart, Kèvin Knoops, Jens Puschhof, Tim I. Breugem, Raimond B. G. Ravelli, et al. 2020. "SARS-CoV-2 Productively Infects Human Gut Enterocytes." *Science*, May, eabc1669. https://doi.org/10.1126/science.abc1669.

Leeman, Kristen T, Patrizia Pessina, Joo-Hyeon Lee, and Carla F Kim. 2019. "Mesenchymal Stem Cells Increase Alveolar Differentiation in Lung Progenitor Organoid Cultures." *Scientific Reports* 9 (1). https://doi.org/10.1038/s41598-019-42819-1.

Lei, Chen, Liu Huiguo, Liu Wei, Liu Jing, Liu Kui, Shang Jin, Deng Yan, and Wei Shuang. 2020. "2019新型冠状病毒肺炎29例临床特征分析." *Chinese Journal of Tuberculosis and Respiratory Diseases* 43 (00): E005–E005. https://doi.org/10.3760/CMA.J.ISSN.1001-0939.2020.0005.

Leng, Zikuan, Rongjia Zhu, Wei Hou, Yingmei Feng, Yanlei Yang, Qin Han, Guangliang Shan, et al. 2020. "Transplantation of ACE2-Mesenchymal Stem Cells Improves the Outcome of Patients with Covid-19 Pneumonia." *Aging and Disease* 11 (2): 216–28. https://doi.org/10.14336/AD.2020.0228.

Li, Mingfen, Xuyong Sun, Xiaocong Kuang, Yan Liao, Haibin Li, and Dianzhong Luo. 2014. "Mesenchymal Stem Cells Suppress CD8+ T Cell-Mediated Activation by Suppressing Natural Killer Group 2, Member D Protein Receptor Expression and Secretion of Prostaglandin E2, Indoleamine 2, 3-Dioxygenase and Transforming Growth Factor-β." *Clinical and Experimental Immunology* 178 (3): 516–24. https://doi.org/10.1111/cei.12423.

Li, Xiaoming, Wenbing Zeng, Xiang Li, Haonan Chen, Linping Shi, Xinghui Li, Hongnian Xiang, et al. 2020. "CT Imaging Changes of Corona Virus Disease 2019(COVID-19): A Multi-Center Study in Southwest China." *Journal of Translational Medicine* 18 (1). https://doi.org/10.1186/s12967-020-02324-w.

Li, Yan, Jun Xu, Weiqing Shi, Cheng Chen, Yan Shao, Limei Zhu, Wei Lu, and Xiao Dong Han. 2016. "Mesenchymal Stromal Cell Treatment Prevents H9N2 Avian Influenza Virus-Induced Acute Lung Injury in Mice." *Stem Cell Research and Therapy* 7 (1): 1–11. https://doi.org/10.1186/s13287-016-0395-z.

Liao, Li, Bingzheng Shi, Heran Chang, Xiaoxia Su, Lichao Zhang, Chunsheng Bi, Yi Shuai, Xiaoyan Du, Zhihong Deng, and Yan Jin.

2017. "Heparin Improves BMSC Cell Therapy: Anticoagulant Treatment by Heparin Improves the Safety and Therapeutic Effect of Bone Marrow-Derived Mesenchymal Stem Cell Cytotherapy." *Theranostics* 7 (1): 106–16. https://doi.org/10.7150/thno.16911.

Liu, Hongyan, Thu Le Trinh, Huijia Dong, Robertson Keith, David Nelson, and Chen Liu. 2012. "Iron Regulator Hepcidin Exhibits Antiviral Activity against Hepatitis C Virus." *PLoS ONE* 7 (10): 3–10. https://doi.org/10.1371/journal.pone.0046631.

Liu, Ling, Hongli He, Airan Liu, Jingyuan Xu, Jibin Han, Qihong Chen, Shuling Hu, et al. 2015. "Therapeutic Effects of Bone Marrow-Derived Mesenchymal Stem Cells in Models of Pulmonary and Extrapulmonary Acute Lung Injury." *Cell Transplantation* 24 (12): 2629–42. https://doi.org/10.3727/096368915X687499.

Liu, Yung Yang, Chi Huei Chiang, Shih Chieh Hung, Chih Feng Chian, Chen Liang Tsai, Wei Chih Chen, and Haibo Zhang. 2017. "Hypoxia-Preconditioned Mesenchymal Stem Cells Ameliorate Ischemia/Reperfusion-Induced Lung Injury." *PLoS ONE* 12 (11): 1–20. https://doi.org/10.1371/journal.pone.0187637.

Lopes-Pacheco, Miquéias, Chiara Robba, Patricia Rieken Macêdo Rocco, and Paolo Pelosi. 2020. "Current Understanding of the Therapeutic Benefits of Mesenchymal Stem Cells in Acute Respiratory Distress Syndrome." *Cell Biology and Toxicology* 36 (1): 83–102. https://doi.org/10.1007/s10565-019-09493-5.

Luz-Crawford, Patricia, Danièle Noël, Ximena Fernandez, Maroun Khoury, Fernando Figueroa, Flavio Carrión, Christian Jorgensen, and Farida Djouad. 2012. "Mesenchymal Stem Cells Repress Th17 Molecular Program through the PD-1 Pathway." *PLoS ONE* 7 (9). https://doi.org/10.1371/journal.pone.0045272.

Masterson, C, M Jerkic, G F Curley, and J G Laffey. 2015. *"Mesenchymal Stromal Cell Therapies: Potential and Pitfalls for ARDS."*

Maumus, Marie, Christian Jorgensen, and Danièle Noël. 2013. "Mesenchymal Stem Cells in Regenerative Medicine Applied to Rheumatic Diseases: Role of Secretome and Exosomes." *Biochimie.* Biochimie. https://doi.org/10.1016/j.biochi.2013.04.017.

Meisel, Roland, Andree Zibert, Maurice Laryea, Ulrich Göbel, Walter Däubener, and Dagmar Dilloo. 2004. "Human Bone Marrow Stromal Cells Inhibit Allogeneic T-Cell Responses by Indoleamine 2,3-Dioxygenase-Mediated Tryptophan Degradation." *Blood* 103 (12): 4619–21. https://doi.org/10.1182/blood-2003-11-3909.

Mendez, Julio J, Mahboobe Ghaedi, Derek Steinbacher, and Laura E Niklason. 2014. "Epithelial Cell Differentiation of Human Mesenchymal Stromal Cells in Decellularized Lung Scaffolds." *Tissue Engineering Part A* 20 (11–12): 1735–46. https://doi.org/10.1089/ten.tea.2013.0647.

Morata-Tarifa, Cynthia, María Del Mar Macías-Sánchez, Antonio Gutiérrez-Pizarraya, and Rosario Sanchez-Pernaute. 2020. "Mesenchymal Stromal Cells for the Prophylaxis and Treatment of Graft-versus-Host Disease - A Meta-Analysis." *Stem Cell Research and Therapy* 11 (1): 64. https://doi.org/10.1186/s13287-020-01592-z.

Nabhan, Ahmad N., Douglas G. Brownfield, Pehr B. Harbury, Mark A. Krasnow, and Tushar J. Desai. 2018. "Single-Cell Wnt Signaling Niches Maintain Stemness of Alveolar Type 2 Cells." *Science* 359 (6380): 1118–23. https://doi.org/10.1126/science.aam6603.

Naji, Abderrahim, Masamitsu Eitoku, Benoit Favier, Frédéric Deschaseaux, Nathalie Rouas-Freiss, and Narufumi Suganuma. 2019. "Biological Functions of Mesenchymal Stem Cells and Clinical Implications." *Cellular and Molecular Life Sciences*. Birkhauser Verlag AG. https://doi.org/10.1007/s00018-019-03125-1.

Nicola, Massimo Di, Carmelo Carlo-Stella, Michele Magni, Marco Milanesi, Paolo D. Longoni, Paola Matteucci, Salvatore Grisanti, and Alessandro M. Gianni. 2002. "Human Bone Marrow Stromal Cells Suppress T-Lymphocyte Proliferation Induced by Cellular or Nonspecific Mitogenic Stimuli." *Blood* 99 (10): 3838–43. https://doi.org/10.1182/blood.V99.10.3838.

Okur, Sibel Çağlar, Sinan Erdoğan, Cansu Subasi Demir, Gülsen Günel, and Erdal Karaöz. 2018. "The Effect of Umbilical Cord-Derived Mesenchymal Stem Cell Transplantation in a Patient with Cerebral

Palsy: A Case Report." *International Journal of Stem Cells* 11 (1): 141–47. https://doi.org/10.15283/ijsc17077.

Omar, Reine El, Jacqueline Beroud, Jean Francois Stoltz, Patrick Menu, Emilie Velot, and Veronique Decot. 2014. "Umbilical Cord Mesenchymal Stem Cells: The New Gold Standard for Mesenchymal Stem Cell-Based Therapies?" *Tissue Engineering - Part B: Reviews*. Mary Ann Liebert Inc. https://doi.org/10.1089/ten.teb.2013.0664.

Oxley, Thomas J., J. Mocco, Shahram Majidi, Christopher P. Kellner, Hazem Shoirah, I. Paul Singh, Reade A. De Leacy, et al. 2020. "Large-Vessel Stroke as a Presenting Feature of Covid-19 in the Young." *New England Journal of Medicine* 382 (20): e60. https://doi.org/10.1056/NEJMc2009787.

Pérez-De-Lis, Marta, Soledad Retamozo, Alejandra Flores-Chávez, Belchin Kostov, Roberto Perez-Alvarez, Pilar Brito-Zerón, and Manuel Ramos-Casals. 2017. "Autoimmune Diseases Induced by Biological Agents. A Review of 12,731 Cases (BIOGEAS Registry)." *Expert Opinion on Drug Safety*. Taylor and Francis Ltd. https://doi.org/10.1080/14740338.2017.1372421.

Pradesh, Uttar, Pradesh Pandit, Deen Dayal, Upadhyaya Pashu, Chikitsa Vigyan, Vishwavidyalaya Evam, Uttar Pradesh, et al. 2020. *Coronavirus Disease 2019 – COVID-19 Kuldeep Dhama*, no. April: 1–75. https://doi.org/10.20944/preprints202003.0001.v2.

Prockop, Darwin J., Malcolm Brenner, Willem E. Fibbe, Edwin Horwitz, Katarina Le Blanc, Donald G. Phinney, Paul J. Simmons, Luc Sensebe, and Armand Keating. 2010. "Defining the Risks of Mesenchymal Stromal Cell Therapy." *Cytotherapy* 12 (5): 576–78. https://doi.org/10.3109/14653249.2010.507330.

Qiu, Guanguan, Guoping Zheng, Menghua Ge, Jiangmei Wang, Ruoqiong Huang, Qiang Shu, and Jianguo Xu. 2018. "Mesenchymal Stem Cell-Derived Extracellular Vesicles Affect Disease Outcomes via Transfer of MicroRNAs." *Stem Cell Research and Therapy* 9: 320.

Quesenberry, Peter J., Jason Aliotta, Maria Chiara Deregibus, and Giovanni Camussi. 2015. "Role of Extracellular RNA-Carrying Vesicles in Cell Differentiation and Reprogramming." *Stem Cell*

Research and Therapy 6 (1). https://doi.org/10.1186/s13287-015-0150-x.

Rizzo, Paola, Francesco Vieceli Dalla Sega, Francesca Fortini, Luisa Marracino, Claudio Rapezzi, and Roberto Ferrari. 2020. "COVID-19 in the Heart and the Lungs: Could We 'Notch' the Inflammatory Storm?" *Basic Research in Cardiology* 115 (3): 1–8. https://doi.org/10.1007/s00395-020-0791-5.

Rosenberg, Jillian, and Jun Huang. 2018. "CD8+ T Cells and NK Cells: Parallel and Complementary Soldiers of Immunotherapy." *Current Opinion in Chemical Engineering*. Elsevier Ltd. https://doi.org/10.1016/j.coche.2017.11.006.

Rozenberg, Ayal, Ayman Rezk, Marie-Noëlle Boivin, Peter J. Darlington, Mukanthu Nyirenda, Rui Li, Farzaneh Jalili, et al. 2016. "Human Mesenchymal Stem Cells Impact Th17 and Th1 Responses Through a Prostaglandin E2 and Myeloid-Dependent Mechanism." *STEM CELLS Translational Medicine* 5 (11): 1506–14. https://doi.org/10.5966/sctm.2015-0243.

Saldaña, Laura, Fátima Bensiamar, Gema Vallés, Francisco J. Mancebo, Eduardo García-Rey, and Nuria Vilaboa. 2019. "Immunoregulatory Potential of Mesenchymal Stem Cells Following Activation by Macrophage-Derived Soluble Factors." *Stem Cell Research and Therapy* 10 (1): 58. https://doi.org/10.1186/s13287-019-1156-6.

Sarvar, Davod Pashoutan, Karim Shamsasenjan, and Parvin Akbarzadehlaleh. 2016. "Mesenchymal Stem Cell-Derived Exosomes: New Opportunity in Cell-Free Therapy." *Advanced Pharmaceutical Bulletin* 6 (3): 293–99. https://doi.org/10.15171/apb.2016.041.

Shabbir, Arsalan, Audrey Cox, Luis Rodriguez-Menocal, Marcela Salgado, and Evangelos Van Badiavas. 2015. "Mesenchymal Stem Cell Exosomes Induce Proliferation and Migration of Normal and Chronic Wound Fibroblasts, and Enhance Angiogenesis in Vitro." *Stem Cells and Development* 24 (14): 1635–47. https://doi.org/10.1089/scd.2014.0316.

Shi, Bingyu, Jingjing Qi, Genhong Yao, Ruihai Feng, Zhuoya Zhang, Dandan Wang, Chen Chen, et al. 2018. "Mesenchymal Stem Cell

Transplantation Ameliorates Sjögren's Syndrome via Suppressing IL-12 Production by Dendritic Cells 11 Medical and Health Sciences 1107 Immunology." *Stem Cell Research and Therapy* 9 (1): 308. https://doi.org/10.1186/s13287-018-1023-x.

Siddiqi, Hasan K., and Mandeep R. Mehra. 2020. "COVID-19 Illness in Native and Immunosuppressed States: A Clinical–Therapeutic Staging Proposal." *Journal of Heart and Lung Transplantation.* Elsevier USA. https://doi.org/10.1016/j.healun.2020.03.012.

Singhal, Tanu. 2020. "A Review of Coronavirus Disease-2019 (COVID-19)." *Indian Journal of Pediatrics* 87 (4): 281–86. https://doi.org/10.1007/s12098-020-03263-6.

Song, Wenfei, Miao Gui, Xinquan Wang, and Ye Xiang. 2018. "Cryo-EM Structure of the SARS Coronavirus Spike Glycoprotein in Complex with Its Host Cell Receptor ACE2." *PLoS Pathogens* 14 (8). https://doi.org/10.1371/journal.ppat.1007236.

Spaggiari, Grazia Maria, Heba Abdelrazik, Flavio Becchetti, and Lorenzo Moretta. 2009. "MSCs Inhibit Monocyte-Derived DC Maturation and Function by Selectively Interfering with the Generation of Immature DCs: Central Role of MSC-Derived Prostaglandin E2." *Blood* 113 (26): 6576–83. https://doi.org/10.1182/blood-2009-02-203943.

Sungnak, Waradon, Ni Huang, Christophe Bécavin, Marijn Berg, and HCA Lung Biological Network. 2020. *SARS-CoV-2 Entry Genes Are Most Highly Expressed in Nasal Goblet and Ciliated Cells within Human Airways.* http://arxiv.org/abs/2003.06122.

Sungnak, Waradon, Ni Huang, Christophe Bécavin, Marijn Berg, Rachel Queen, Monika Litvinukova, Carlos Talavera-López, et al. 2020. "SARS-CoV-2 Entry Factors Are Highly Expressed in Nasal Epithelial Cells Together with Innate Immune Genes." *Nature Medicine*, no. Mdc. https://doi.org/10.1038/s41591-020-0868-6.

Sutton, Morgan T, David Fletcher, Nicole Episalla, Lauren Auster, Michael Folz, Varun Roy, Rolf Van Heeckeren, Donald P Lennon, Arnold I Caplan, and Tracey L Bonfield. 2017. "Mesenchymal Stem Cell Soluble Mediators and Cystic Fibrosis." *Journal of Stem Cell*

Research & Therapy 07 (09). https://doi.org/10.4172/2157-7633.1000400.

Tang, Ning, Huan Bai, Xing Chen, Jiale Gong, Dengju Li, and Ziyong Sun. 2020. "Anticoagulant Treatment Is Associated with Decreased Mortality in Severe Coronavirus Disease 2019 Patients with Coagulopathy." *Journal of Thrombosis and Haemostasis*, no. March: 1094–99. https://doi.org/10.1111/jth.14817.

Tang, Ning, Dengju Li, Xiong Wang, and Ziyong Sun. 2020. "Abnormal Coagulation Parameters Are Associated with Poor Prognosis in Patients with Novel Coronavirus Pneumonia." *Journal of Thrombosis and Haemostasis* 18 (4): 844–47. https://doi.org/10.1111/jth.14768.

Tripathi, Shweta, Guangshun Wang, Mitchell White, Li Qi, Jeffery Taubenberger, and Kevan L. Hartshorn. 2015. "Antiviral Activity of the Human Cathelicidin, LL-37, and Derived Peptides on Seasonal and Pandemic Influenza A Viruses." *PLoS ONE* 10 (4): 1–17. https://doi.org/10.1371/journal.pone.0124706.

Turner, Anthony J. 2015. "ACE2 Cell Biology, Regulation, and Physiological Functions." In *The Protective Arm of the Renin Angiotensin System (RAS): Functional Aspects and Therapeutic Implications*, 185–89. Elsevier Inc. https://doi.org/10.1016/B978-0-12-801364-9.00025-0.

Tyndall, Alan. 2011. "Successes and Failures of Stem Cell Transplantation in Autoimmune Diseases." *Hematology / the Education Program of the American Society of Hematology. American Society of Hematology. Education Program.* American Society of Hematology. https://doi.org/10.1182/asheducation-2011.1.280.

Uccelli, Antonio, Lorenzo Moretta, and Vito Pistoia. 2006. "Immunoregulatory Function of Mesenchymal Stem Cells." *European Journal of Immunology* 36 (10): 2566–73. https://doi.org/10.1002/eji.200636416.

———. 2008. "Mesenchymal Stem Cells in Health and Disease." *Nature Reviews Immunology* 8 (9): 726–36. https://doi.org/10.1038/nri2395.

Udugama, Buddhisha, Pranav Kadhiresan, Hannah N. Kozlowski, Ayden Malekjahani, Matthew Osborne, Vanessa Y.C. Li, Hongmin Chen,

Samira Mubareka, Jonathan B. Gubbay, and Warren C.W. Chan. 2020. "Diagnosing COVID-19: The Disease and Tools for Detection." *ACS Nano*. NLM (Medline). https://doi.org/10.1021/acsnano.0c02624.

Viboud, Cécile, and Lone Simonsen. 2012. "Global Mortality of 2009 Pandemic Influenza A H1N1." *The Lancet Infectious Diseases*. Elsevier. https://doi.org/10.1016/S1473-3099(12)70152-4.

Vieira Paladino, Fernanda, Luiz Roberto Sardinha, Carla Azevedo Piccinato, and Anna Carla Goldberg. 2017. *Intrinsic Variability Present in Wharton's Jelly Mesenchymal Stem Cells and T Cell Responses May Impact Cell Therapy*. https://doi.org/10.1155/2017/8492797.

Wang, Baohong, Mingfei Yao, Longxian Lv, Zongxin Ling, and Lanjuan Li. 2017. "The Human Microbiota in Health and Disease." *Engineering*. Elsevier Ltd. https://doi.org/10.1016/J.ENG.2017.01.008.

Wang, Dandan, Xuebing Feng, Lin Lu, Joanne E. Konkel, Huayong Zhang, Zhiyong Chen, Xia Li, et al. 2014. "A CD8 T Cell/Indoleamine 2,3-Dioxygenase Axis Is Required for Mesenchymal Stem Cell Suppression of Human Systemic Lupus Erythematosus." *Arthritis and Rheumatology* 66 (8): 2234–45. https://doi.org/10.1002/art.38674.

Wang, Dandan, Jing Li, Yu Zhang, Miaojia Zhang, Jinyun Chen, Xia Li, Xiang Hu, Shu Jiang, Songtao Shi, and Lingyun Sun. 2014. "Umbilical Cord Mesenchymal Stem Cell Transplantation in Active and Refractory Systemic Lupus Erythematosus: A Multicenter Clinical Study." *Arthritis Research and Therapy* 16 (2). https://doi.org/10.1186/ar4520.

Wang, Dandan, and Lingyun Sun. 2018. "Systemic Lupus Erythematosus." In *A Roadmap to Nonhematopoietic Stem Cell-Based Therapeutics: From the Bench to the Clinic*, 143–72. Elsevier. https://doi.org/10.1016/B978-0-12-811920-4.00007-0.

Wang, Janice, Negin Hajizadeh, Ernest E. Moore, Robert C. McIntyre, Peter K. Moore, Livia A. Veress, Michael B. Yaffe, Hunter B. Moore, and Christopher D. Barrett. 2020. "Tissue Plasminogen Activator (TPA) Treatment for COVID-19 Associated Acute Respiratory

Distress Syndrome (ARDS): A Case Series." *Journal of Thrombosis and Haemostasis : JTH*. https://doi.org/10.1111/jth.14828.

Wang, Lifeng, Fu Sheng Wang, and M. Eric Gershwin. 2015. "Human Autoimmune Diseases: A Comprehensive Update." *Journal of Internal Medicine*. Blackwell Publishing Ltd. https://doi.org/10.1111/joim.12395.

Wang, Wenling, Yanli Xu, Ruqin Gao, Roujian Lu, Kai Han, Guizhen Wu, and Wenjie Tan. 2020. "Detection of SARS-CoV-2 in Different Types of Clinical Specimens." *JAMA - Journal of the American Medical Association*, 5–6. https://doi.org/10.1001/jama.2020.3786.

WHO. 2020a. "Novel Coronavirus(2019-NCoV) Situation Report – 22."

———. 2020b. *Coronavirus Disease 2019 (COVID-19) Situation Report – 51*.

Wilson, Jennifer G., Kathleen D. Liu, Hanjing Zhuo, Lizette Caballero, Melanie McMillan, Xiaohui Fang, Katherine Cosgrove, et al. 2015. "Mesenchymal Stem (Stromal) Cells for Treatment of ARDS: A Phase 1 Clinical Trial." *The Lancet Respiratory Medicine* 3 (1): 24–32. https://doi.org/10.1016/S2213-2600(14)70291-7.

Worthington, Erin N., and James S. Hagood. 2020. "Therapeutic Use of Extracellular Vesicles for Acute and Chronic Lung Disease." *International Journal of Molecular Sciences* 21 (7). https://doi.org/10.3390/ijms21072318.

Wu, Aiping, Yousong Peng, Baoying Huang, Xiao Ding, Xianyue Wang, Peihua Niu, Jing Meng, et al. 2020. "Genome Composition and Divergence of the Novel Coronavirus (2019-NCoV) Originating in China." *Cell Host and Microbe* 27 (3): 325–28. https://doi.org/10.1016/j.chom.2020.02.001.

Xiang, Bingyu, Lu Chen, Xiaojun Wang, Yongjia Zhao, Yanling Wang, and Charlie Xiang. 2017. "Transplantation of Menstrual Blood-Derived Mesenchymal Stem Cells Promotes the Repair of Lps-Induced Acute Lung Injury." *International Journal of Molecular Sciences* 18 (4). https://doi.org/10.3390/ijms18040689.

Xie, Chunbao, Lingxi Jiang, Guo Huang, Hong Pu, Bo Gong, He Lin, Shi Ma, et al. 2020. "Comparison of Different Samples for 2019 Novel Coronavirus Detection by Nucleic Acid Amplification Tests." *International Journal of Infectious Diseases* 93: 264–67. https://doi.org/10.1016/j.ijid.2020.02.050.

Xu, Hao, Liang Zhong, Jiaxin Deng, Jiakuan Peng, Hongxia Dan, Xin Zeng, Taiwen Li, and Qianming Chen. 2020. "High Expression of ACE2 Receptor of 2019-NCoV on the Epithelial Cells of Oral Mucosa." *International Journal of Oral Science* 12 (1): 1–5. https://doi.org/10.1038/s41368-020-0074-x.

Ye, Qing, Bili Wang, and Jianhua Mao. 2020. "The Pathogenesis and Treatment of the 'Cytokine Storm' in COVID-19." *Journal of Infection* 80: 607–13. https://doi.org/10.1016/j.jinf.2020.03.037.

Yilmaz, Sema, Nihal Inandiklioglu, Dincer Yildizdas, Cansu Subasi, Arbil Acikalin, Yurdun Kuyucu, Ibrahim Bayram, et al. 2013. "Mesenchymal Stem Cell: Does It Work in an Experimental Model with Acute Respiratory Distress Syndrome?" *Stem Cell Reviews and Reports* 9 (1): 80–92. https://doi.org/10.1007/s12015-012-9395-2.

Yin, Kan, Shihua Wang, and Robert Chunhua Zhao. 2019. "Exosomes from Mesenchymal Stem/Stromal Cells: A New Therapeutic Paradigm." *Biomarker Research*. BioMed Central Ltd. https://doi.org/10.1186/s40364-019-0159-x.

Zaim, Sevim, Jun Heng Chong, Vissagan Sankaranarayanan, and Amer Harky. 2020. "COVID-19 and Multiorgan Response." *Current Problems in Cardiology*. Mosby Inc. https://doi.org/10.1016/j.cpcardiol.2020.100618.

Zhang, Bin, Mei Wang, Aihua Gong, Xu Zhang, Xiaodan Wu, Yanhua Zhu, Hui Shi, et al. 2015. "HucMSc-Exosome Mediated-Wnt4 Signaling Is Required for Cutaneous Wound Healing." *Stem Cells* 33 (7): 2158–68. https://doi.org/10.1002/stem.1771.

Zhang, Chi, Zhao Wu, Jia Wen Li, Hong Zhao, and Gui Qiang Wang. 2020. "The Cytokine Release Syndrome (CRS) of Severe COVID-19 and Interleukin-6 Receptor (IL-6R) Antagonist Tocilizumab May Be the Key to Reduce the Mortality." *International Journal of Antimicrobial Agents* 55 (5): 105954. https://doi.org/10.1016/j.ijantimicag.2020.105954.

Zhang, Jieyuan, Junjie Guan, Xin Niu, Guowen Hu, Shangchun Guo, Qing Li, Zongping Xie, Changqing Zhang, and Yang Wang. 2015. "Exosomes Released from Human Induced Pluripotent Stem Cells-Derived MSCs Facilitate Cutaneous Wound Healing by Promoting Collagen Synthesis and Angiogenesis." *Journal of Translational Medicine* 13 (1). https://doi.org/10.1186/s12967-015-0417-0.

Zhang, Su fen, Jiu ling Tuo, Xu bin Huang, Xun Zhu, Ding mei Zhang, Kai Zhou, Lei Yuan, et al. 2018. "Epidemiology Characteristics of Human Coronaviruses in Patients with Respiratory Infection Symptoms and Phylogenetic Analysis of HCoV-OC43 during 2010-2015 in Guangzhou." *PLoS ONE* 13 (1): 1–20. https://doi.org/10.1371/journal.pone.0191789.

Zhang, Xiwen, Jianxiao Chen, Ming Xue, Yuying Tang, Jingyuan Xu, Ling Liu, Yingzi Huang, Yi Yang, Haibo Qiu, and Fengmei Guo. 2019. "Overexpressing P130/E2F4 in Mesenchymal Stem Cells Facilitates the Repair of Injured Alveolar Epithelial Cells in LPS-Induced ARDS Mice." *Stem Cell Research & Therapy* 10 (1). https://doi.org/10.1186/s13287-019-1169-1.

Zheng, Guoping, Lanfang Huang, Haijiang Tong, Qiang Shu, Yaoqin Hu, Menghua Ge, Keqin Deng, et al. 2014. "Treatment of Acute Respiratory Distress Syndrome with Allogeneic Adipose-Derived Mesenchymal Stem Cells: A Randomized, Placebo-Controlled Pilot Study." *Respiratory Research* 15 (1): 39. https://doi.org/10.1186/1465-9921-15-39.

Zhou, Fei, Ting Yu, Ronghui Du, Guohui Fan, Ying Liu, Zhibo Liu, Jie Xiang, et al. 2020. "Clinical Course and Risk Factors for Mortality of Adult Inpatients with COVID-19 in Wuhan, China: A Retrospective Cohort Study." *The Lancet* 395 (10229): 1054–62. https://doi.org/10.1016/S0140-6736(20)30566-3.

Zulueta, Aida, Michela Colombo, Valeria Peli, Monica Falleni, Delfina Tosi, Mario Ricciardi, Alessandro Baisi, Gaetano Bulfamante, Raffaella Chiaramonte, and Anna Caretti. 2018. "Lung Mesenchymal Stem Cells-Derived Extracellular Vesicles Attenuate the Inflammatory Profile of Cystic Fibrosis Epithelial Cells." *Cellular Signalling* 51: 110–18. https://doi.org/https://doi.org/10.1016/j.cellsig.2018.07.015.

In: Human Mesenchymal Stem Cells
Editor: Mitchell Khan

ISBN: 978-1-53619-155-4
© 2021 Nova Science Publishers, Inc.

Chapter 2

ACTIVATED MESENCHYMAL STEM CELLS FOR STROKE REPAIR

*Ravi Prakash, Santosh Kumar Yadav,
Abu Junaid Siddiqui, Neha Kumari, Mohsin Ali Khan
and Syed Shadab Raza*[*]

Laboratory for Stem Cell and Restorative Neurology,
Era's Lucknow Medical College and Hospital,
Era University, Lucknow, India

ABSTRACT

Cerebral stroke is a severe health concern. Stem Cell treatment has shown recovery in animal models of stroke, indicating an improved regenerative and repair potential. Though stem cells are still being used in clinical trials, there is no evidence that they enhance recovery in ischemic stroke patients. Nevertheless, the multipotent mesenchymal stem has widely been explored for stroke recovery. Low in vivo MSC survival, intrinsic differences between MSC, sources and donor variability, and

---

[*] Corresponding Author's E-mail: drshadab@erauniversity.in.

variability of culturing protocols have been described as few limitations in the field. Strikingly, post-transplantation MSCs could be regulated by the locally regulated environment, indicating that restorative variability could be managed by selecting a priming regimen to rectify a given pathology precisely. In stem cell treatment, a new area of research is the preconditioning of cells before transplantation. As no effective treatment for stroke recovery is available, and owing to the fact that MSCs could be customized, we propose 'Activated MSC' as a therapeutic alternative to tackling ischemic stroke. Therefore, the activation of MSCs by cytokines, growth factors, hypoxia, pharmacological drugs, etc., could be a novel approach to improving stroke patients' responses to receiving MSCs.

ABBREVIATIONS

Ang-2	Angiotensin 2
Ang-4	Angiogenin 4
BDNF	Brain derived nerve growth factor
BM-MSCs	Bone marrow derived mesenchymal stem cells
CCL2	C-C Motif Chemokine Ligand 2
CXCR	CXC chemokine receptor;
Cyt c	Cytochrome c
COX-2	Cyclooxygenase-2
CCR2	C-C Motif Chemokine Receptor 2
FGF	Fibroblast Growth Factors
FHG	Fish Growth Harmone
GDNF	Glial Cell Line-Derived Neurotrophic Factor
HIF1α	Hypoxiainducible factor1 alpha
Hsp	Heat shock protein
HGF	Hepatocyte Growth Factor
IL-10	Interleukin 10
IL-1β	Interleukin-1β
IFN-γ	Interferon Gamma;
ICAM-1	Intercellular Adhesion Molecule 1
ICAM-4	Intercellular Adhesion Molecule 4
IDO	Indoleamine-pyrrole 2,3-dioxygenase
LPS	Lipopolysaccharide

MHC	Major Histocompatibility
MMP1	Matrix Metalloproteinase-1
MMP13	Matrix Metallopeptidase 13
MMP	Matrix mettallopeptidase
NCX1	Sodium–calcium exchanger1
NF-Kb	Nuclear Factor kappa B
PDGF	Platelet-Derived Growth Factor
PlGF	Placental Growth Factor
SDF1	Stromal derived factor1
STAT	Signal Transducer and Activator of Transcription Proteins
TNF-α	Tumor Necrosis Factor alpha
TGF-b	Transforming Growth Factor beta
TLR-3	Toll-Like Receptor 3
VEGF	Vascular endothelial growth factor
VCAM-1	Vascular cell adhesion molecule 1
VEGF-4	Vascular Endothelial Growth Factor 4

INTRODUCTION

Stroke is the world's second leading cause of mortality and the first leading cause of long-term disability (Musuka et al., 2015). The lifetime risk for stroke is estimated between 8% and 10%. It takes around 6.7 million lives per annum. A current estimate accessed nearly 33 million living with stroke to live disability-adjusted life years (Cunningham et al., 2020). Definition of stroke incorporates a focal neurological deficit resulting from disruption of the cerebral blood supply. It is characterized by progressive dysfunction of neurons, glial cells, and their networks, leading to neuronal degeneration. Ischemic stroke is the most common stroke type (approximately 85% of all strokes), while primary intracerebral bleeding (hemorrhagic stroke) accounts for the remainder. Ischemia leads to a reduction in blood flow adequate to alter normal cellular function. Brain tissue is extremely fragile to ischemia, such that even a brief ischemic period to neurons can induce a complex series of events that ultimately may lead to cellular death.

During the last few decades, there has been a significant advancement in understanding and treating stroke patients (Petrovic-Djergovic et al., 2016). While several clinical trials have been performed in patients with ischemic stroke, none of these have promised preventive efficiency except for intravenous alteplase (tPA) thrombolysis. The clinical studies show that intravenous tPA is effective and safe when given under controlled conditions within 3-4 hours after stroke onset. However, after this time point, tPA administration remains controversial (Blakeley and Llinas, 2007; Marder et al., 2010). Hence, there is a considerable demand for alternative therapeutic approaches to enhance stroke patients' recovery. Possible alternative treatments include regenerative therapies, including stem cells. Many cell types isolated from numerous sources are available today (Raza et al., 2018). Mesenchymal Stem Cells (MSC) is one of those. MSCs are pluripotent in nature with capabilities for self-renewal and multidirectional differentiation. The therapeutic potential of MSC is attributed to their ability to home to the specific target, to undergo lineage-specific differentiation and integration, to modulate the immune system, and the ability to release bioactive factors (Raza et al., 2018). In the above lines, MSCs may serve as a promising candidate for cell replacement therapy in regenerative medicine (Ullah et al., 2015; Squillaro et al., 2016). However, apart from the years of success, a few challenges encountered by MSCs for successful transplantation at clinics include heterogeneous treatment response, optimal doses, optimal timing, optimal route, low viability, source-specific immunomodulatory response (Raza et al., 2018; Surugiu et al., 2019). In MSCs based therapy and tissue remodeling, the microenvironments of both the donors and recipients play an imperative role in the determination of regenerative efficiency of transplanted cells (Ko et al., 2015; Sui et al., 2017), indicating that the therapeutic potential of MSCs can be managed and is an attractive area for investigation. For instance, the resting or naïve MSCs do not demonstrate immunomodulatory or homing potentials; however, only when they are exposed to a stimuli milieu (Carvalho et al., 2019; Ren et al., 2008). This indicates that the modulatory activity is not constitutively expressed by MSCs but is set on by the process of 'priming or licensing' to be obtained. Pre-conditioning has evolved as a chief therapeutic strategy for various regenerative studies (Raza et al., 2020), including cerebral injuries (Nichols et al., 2013; Pan et al., 2020) in laboratories. However, translating the success of laboratory studies to human cerebral protection will dictate laboratory studies' success. In the above lines, the modulation of biological, biochemical, and biophysical properties by external factors could influence the fate of the transplanted cells, thereby enabling an efficient strategy to enhance therapeutic potential (Huang et al., 2015; Nava et al., 2012). Here, we summarized the effect of pre-conditioning the MSCs on their curative potential that may contribute to developing treatments for neurological disorders.

MSCs AND STROKE RECOVERY

MSCs, initially known as Mesenchymal Stromal Cells, are fibroblast-like non-hematopoietic cells, according to the International Society for Stem Cell Research (ISSCR). The term "mesenchymal stem cell" is often used for cells that meet the following standards: adherence to tissue culture plastic; ability to differentiation into osteoclasts, adipocytes, and chondroblasts *in vitro*; expression of CD73, CD90, and CD105 surface markers; and negative for CD34, CD45, CD14 or CD11b, C79a or CD19 and HLA-DR (Dominici et al., 2006). Primarily, MSCs were assumed to be found in bone marrow, though currently, they are isolated from different organs and tissues (Meirelles et al., 2006) such as the placenta, umbilical cord (Romanov et al., 2003), Wharton's jelly (Wang et al., 2004), adipose tissue (Zuk et al., 2002), the dental pulp (Gronthos et al., 2002), lungs, liver, spleen, and brain (isn't Anker et al., 2003; Jiang et al., 2002).

The number of MSC based clinical trials has increased significantly in the last decade. To date, the number of clinical trials employing MSCs is 949 as per the official database of the US National Institutes of Health (https://www.clinicaltrials.gov/ct2/results?cond=Mesenchymal+Stem+Cell

s&term=&cntry=&state=&city=&dist=). Currently, several clinical trials are being conducted to examine the capability of MSCs for cell-based therapy in the treatment of a vast variety of neurological disorders (https://www.clinicaltrials.gov/ct2/results?cond=Mesenchymal+Stem+Cells+&term=Neurological+Disorders&cntry=&state=&city=&dist=) including stroke (https://www.clinicaltrials.gov/ct2/results?cond=Mesenchymal+Stem+Cells+&term=Stroke+&cntry=&state=&city=&dist=). The results from Phase I/II trials advocate the safety and feasibility of MSCs in the treatment of ischemic stroke (Bhasin et al., 2017; Steinberg et al., 2016). However, most of the trials have either indicated very low efficacy of MSCs or none at clinics. For example, the study by Bang et al., using BM-MSCs as transplant revealed that the intravenous infusion of 1 × 10^8 BM-MSCs twice within nine weeks after the ischemic stroke had no adverse effect, including death, while an improvement in Barthel's Index in patients receiving BM-MSCs was observed compared with control during the first year of observation (Bang et al., 2005).

Exogenous stem cell therapy has been recognized as a potential reparative approach for growing regeneration for stroke survivors. Mounting evidence suggests that the primary source of stem cells used in the bulk of both pre-clinical and clinical stroke trials is MSCs (Chen et al., 2001a; Deng et al., 2019; Jaillard et al., 2020; Zin'kova et al., 2007). Due to the ease of harvesting and maintenance in the lab, ease of ethical issues, etc., the use of MSCs is justified compared to other cell types (adult or embryonic). The brain controls bone marrow through pathways that include the sympathetic nervous system and systemic inflammation. The process is known to get activated after a stroke (Courties et al., 2015; Wolf and Ley, 2015), indicating a biological or physiological connection with stroke patients' bone marrow-derived MSCs. However, for the sub-acute phase, MSC transplants have been shown to impede oxidative stress, inflammation, mitochondrial impairment, apoptosis, and early secondary cell death responses associated with stroke (Kasahara et al., 2016). Through the virtue of their secretary behavior, MSCs induce angiogenesis, neurogenesis, and neurite outgrowth in the surrounding tissues of the core region of the infarct (Vu et al., 2014; Xin et al., 2013; Dulamea, 2015). On these lines, MSCs treatment in the chronic phase has been demonstrated to trigger brain remodeling (Horie et al., 2011). Further, MSCs preconditioned with chemo-attractants or delivered synergistically with chemo-attractants can guide the migration of MSCs from the periphery to the brain (Reyes et al., 2015). Previous studies have reported the migration of transplanted MSCs to areas of cerebral infarct and mediated tissue healing (Lee et al., 2015). Nevertheless, it is imperative to determine the route, dose, and timing of MSC transplantation for stroke therapy's success. The intracerebral administration yields the most benefits to the ischemic brain; it is more invasive than intravascular cell transplantation. Therefore, the minimally invasive intravenous or intra-arterial delivery remains the favored option for the sub-acute phase in a previously injured brain shaped by the main ischemic insult. Stereotactic injection and intra-ventricular injection are the main intracranial procedures (Toyoshima et al., 2017). The stereotactic injection helps directly transport the MSCs to the region of infarction. However, to hit the target location, i.e., penumbra in the infarct brain, the technique demands high precision (Jin et al., 2005). The advantage of employing intra-ventricular infusion is that the injected cells could reach broader cortical areas. However, the therapeutic effect depends on the number of cells in such cases. Therefore, it is reasonable to say that intravascular delivery is the safest and most effective mode for more prominent lesions, although a higher spread is required. Nevertheless, the therapeutic benefit could be diluted over a broader volume. Similarly, the timing of transplantation is advantageous in case of stroke injury. For instance, in a permanent model of rodent middle cerebral artery occlusion (MCAO), very early transplantation of human MSCs produced excellent neurological recovery and reduced infarction volume that led to more remarkable neurological improvement post-ischemic stroke (Nam et al., 2015). It is worth mentioning that the different goals for MSC treatment are provided by the various phases of the pathological ischemic procedure. In the early stage, MSC infusion can impede inflammatory response, regulate toxicity in the complex environment, and reduce peri-infarct region injury. Late cell transplantation (2-3 weeks after ischemia) can modulate reparative processes in favor of angiogenesis and neurogenesis (Muir, 2017).

ACTIVATION OF MSCS

To optimize the therapeutic effects of MSCs, researchers have attempted several ways to refine the technique, including the source of the MSCs, mode of delivery, the timing of the transplant, and dosage, i.e., the number of cells (Friis et al., 2011; Jiang et al., 2011; Lehrke et al., 2006). However, due to the harsh microenvironment of the ischemic region, such optimizations have failed to demonstrate satisfactory progress in the poor retention and viability of transplanted cells, which necessitated the development of innovative techniques to improve the MSC survival and functioning. In this regard, the manipulation of MSCs by multiple methods to enhance the clinical outcomes has now been practiced using various chemicals, pharmacological drugs, and physical and biological agents (Luo et al., 2019). For instance, the MSCs could be trained to face the unfavorable condition with hypoxia and various genetic and epigenetic manipulations to boost their transplantation roles, effectiveness, and protection.

In this chapter, we will deal with a few approaches that have been put forwarded to enhance the effectiveness, endurance, and therapeutic efficiency of MSCs. Approaches leading to the activation of MSCs through a variety of stimuli to bolster the cells' therapeutic potential is depicted in Figure 1. The mentioned approaches referred to as "licensing and priming" include exposing the MSCs with physical treatments (such as hypoxia and heat shock), pharmacological agents (e.g., LPS, VPA) (Price et al., 1994; Tasaki et al., 1997), cytokines (IL-1α, IL-1β) (Redondo-Castro et al., 2017, Magne et al., 2020), cells engineered to over-express growth factors (e.g., BDNF, HGF) (Kurozumi et al., 2004; Zhao et al., 2006), hypoxia pre-treatment (Beegle et al., 2015; Du et al., 2014), modified culture conditions (e.g., serum deprivation) (Oskowitz et al., 2011], chemicals (2,4-

dinitrophenol (DNP), dimethyloxalylglycine (DMOG)) (Loganathan et al., 2011; Rey et al., 2011) or other various molecules— *in vitro* before transplantation into the brain.

Table 1. Various activation approaches for MSCs empowerment and their advantages and disadvantages

S. No.	Activation method	Pros	Cons
1.	Cytokine and Growth factors	1. Cytokines may enhance anti-inflammatory MSC phenotype 2. Growth factors can increase the rate of survival, homing, and migration	1. The priming of MSCs by proinflammatory cytokines MHC class I and class II level post-implantation in vivo 2. Growth factors may affect the differentiation and integration capacity of MSCs
2.	Hypoxia	1. Hypoxia preconditioning can enhance homing capacity and migratory potential. 2. Hypoxia precondition can also enhance neurogenic and vaculogenic potential of MSCs	1. Buildup of reactive oxygen species 2. Could lead to induce oxidative stress in the cells 3. Putative negative effect on differentiation capacity of preconditioned cells may be evident
3.	Pharmacological Activations	1. Increased production of immunoregulatory molecules 2. Lead to an enhanced survival, homing, and migration	1. Influences MSC differentiation and integration capacities
4.	3 D spheroid cultures	1. Enhanced stemness, thereby, differentiation and integration potential	1. The variability in size may lead to the necrotic core 2. Interfere with the effectiveness of MSCs post transplantation 3. Size variability may hinder effective *in vivo* implantation 4. Limited diffusion of nutrients and oxygen in 3D spheroid core, depending on size

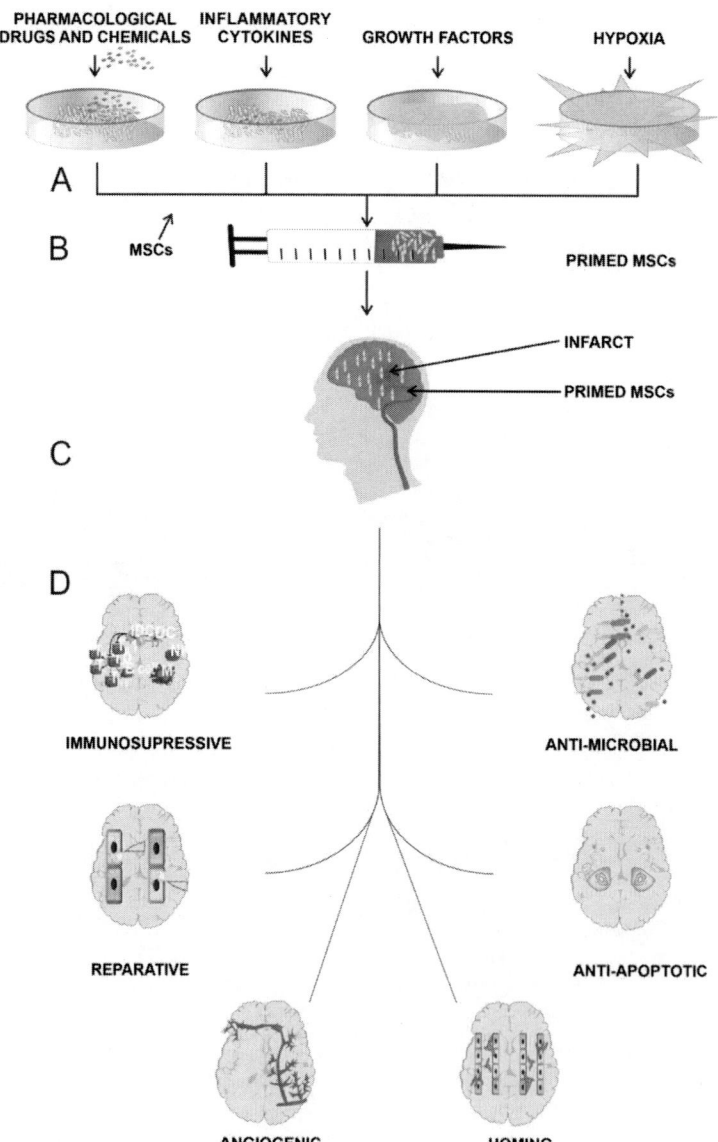

Figure 1. Representative figure showing the priming approaches to improve MSCs therapeutic success for the treatment of stroke patients. (a) Represent pre-conditioning approaches (priming by pharmacological drugs and chemicals, cytokines, growth factors, and hypoxia pre-conditioning, (b) Primed MSCs before transplantation, (c) 'Primed MSCs' post-infusion in the infarcted brain, (d) Therapeutic effect exerted by 'Activated MSCs' (immunosuppression, reparative, angiogenesis, homing, anti-apoptotic, and anti-microbial activities).

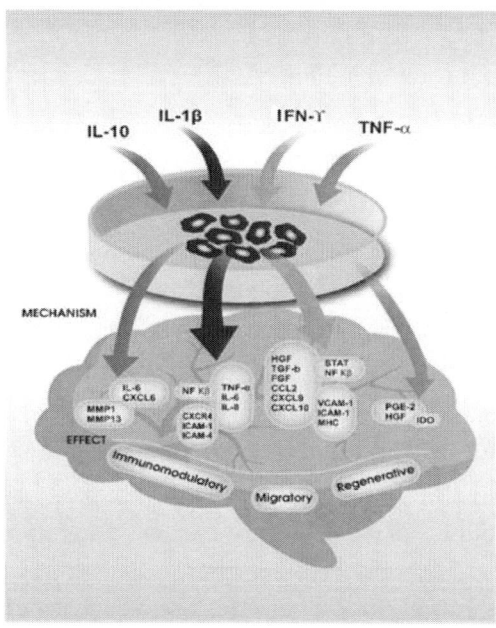

Figure 2. An overview of the production of 'Activated MSCs' for the treatment of stroke. Preconditioning with cytokine or chemokines improves the therapeutic efficacy of MSCs. Arrows and boxes link preconditioning agents and the respectively triggered mechanisms in the center of the stroke brain. The priming effects on MSC, i.e., immunomodulation, migration, and regeneration activated by the cytokines, are indicated in the horizontal boxes at the water-shaded brain picture's lower interfaces (partially adapted from Noronha et al., 2019).

Activation perspectives and MSCs robustness are at present under development and investigation, at present. *In vitro* treatment of MSCs with the afore-mentioned stimulus, before transplantation, may intensify the expression of fitness markers and stimulate exceptional therapeutic outcomes. In our view, the MSCs activation could be achieved through but not limited to (i) Licensing with inflammatory cytokines or mediators, (ii) priming with chemokines, (iii) pre-treatment with hypoxia, (iv) exposure with pharmacological drugs and chemical agents, (v) MSC priming with biomaterials, (vi) cell-engineering for genetic or epigenetic modifications, (vii) Activation by spheroid culture, and (viii) the timing of MSCs engraftment and engagement. The advantages and non-advantages associated with the activation of MSCs are summarized in Table 1. A couple of approaches for obtaining 'Activated MSCs' are discussed above.

MSCs AND CYTOKINES

Activation of MSCs by cytokines or chemokines is one of the techniques implemented in the specifically designed or engineered setting during the *ex vivo* expansion of MSCs with different physical or chemical parameters/factor(s) to activate the adaptive response and boost the capacity of MSCs to work under unfavorable conditions (Saparov et al., 2016; Zhao et al., 2020). Thus, the proliferative, secretary, migratory, and differentiation abilities of MSCs can be enhanced significantly by adequate pretreatment of MSCs by cytokines and chemokines before infusion, favoring beneficial effects in vivo (Hu and Li, 2018) (Figure 2). MSCs derived from various sources and different organisms (human, equine, murine) have successfully been preconditioned by these techniques. These *in vitro* preconditioning approaches are selective and seek to enhance the effectiveness of MSCs in terms of homing, retention, viability, and therapeutic potential. Table 2 summarizes the use of various cytokines or growth factors to empower MSCs in ischemic stroke treatment.

The inflammatory reaction is one of the significant characteristics of stroke pathophysiology. The inflammation is a double-edged sword as it is essential for tissue repair and also a vital player for secondary brain injury after stroke (Barone and Feyerstein, 1999; Yan et al., 2015). Since it plays a significant role in neuronal damage, it becomes a prime target for developing new stroke therapies. As MSC is a potent immunomodulator, it is preferably applied to suppress inflammatory responses in clinical trials (Rawat et al., 2019). To strengthen this property, different research groups have primed the MSCs with a variety of mediators, including but not limited to IL-1 (Redondo-Castro et al., 2017), TNF-a (Aktas et al., 2017; Lu et al., 2017), IFN-γ (Maumus et al., 2013; Sivanathan et al., 2014) or combinations of these (Barrachina et al., 2017). The primed MSCs secreted elevated levels of immunomodulatory mediators, including prostaglandin E2 (PGE2), IL-6, and granulocyte colony-stimulating factor (G-CSF) in response to the priming stimuli, hence up-regulating the expression of the reparative molecule.

Interleukins, known modulators of cellular activity, play a crucial role in cell growth, differentiation, motility, and stimulating immune responses. Based on the primary effect, we can categorize inflammatory interleukins into pro-and anti-inflammatory interleukins. In the ischemic brain, inflammation is vitally regulated by inflammatory cytokines such as interleukin-1 (IL-1) and tumor necrosis factor-alpha (TNF-α), synthesized by microglia, astrocytes, and endothelial cells. Since the preconditioning can boost the secretary profile of MSCs, therefore, the priming of MSCs with specific inflammatory interleukins can train the cells to transform into an anti-inflammatory phenotype more effectively in response to future lethal inflammatory microenvironments (Redondo-Castro et al., 2017). For example, the preconditioning of MSC with IL-1 could elicit the release of the trophic factor G-CSF. Likewise, the supply of IL-1-primed MSC conditioned media to inflamed microglial cells can result in a reduction in the secretion of inflammatory markers (IL-6, G-CSF, and TNF-α), and an enhancement in the microglial-derived anti-inflammatory mediator cytokine IL-10 could also be observed. Taking the clues from these data, the beneficial effects of IL-1α primed MSC conditioned medium in an *in vivo* mouse model with middle cerebral artery occlusion were validated. The authors observed that the subcutaneous administration of IL-1α-primed MSCs conditioned medium at the time of reperfusion has a neuroprotective effect leading to ~30% reduction in lesion volume. Further, the delayed administration (24 h post-ischemia) accelerated recovery independent of neuroprotection (Cunningham et al., 2020). However, pro-inflammatory priming can lead to increase immunogenicity (Barrachina et al., 2017). Immunogenicity could be a determining factor in the success of potential cell therapies. Brief priming times can be used to limit this undesirable effect with low doses of pro-inflammatory mediators.

Table 2. Cytokines and growth factors mediated pre-conditioning

S no	Source of priming	Cell type	Effect	References
1	IL-1α	hMSCs	IL-1α activated hMSC receiving group showed: 1. 30% reduction in lesion volume at 48 h 2. Improvement in body mass gain 3. Improvement in 28-point neurological score 4. Improvement in nest building property	Cunningham et al., 2020
2	IL-10 (gene modification)	hMSCs	IL-10 gene modified hMSCs showed: 1. A significant reduced infarct volume 2. Improvement in motor function 3. Suppression of neuronal degeneration 4. Enhanced neuronal viability 5. Improved survival of engrafted MSCs	Nakajima et al., 2017
3	BNDF (gene modification)	hMSCs	BNDF expressing hMSCs receiving group showed: 1. A significant functional recovery 2. Reduced apoptotic cell death	Kurozumi et al., 2004
4	Angiopoietin-1 and VEGF (gene modification)	hMSCs	Angiopoietin-1 and VEGF expressing hMSCs receiving group showed: 1. A structural-functional recovery 2. Enhanced therapeutic effects	Toyama et al., 2009
5	GDNF (gene modification)	hMSCs	GDNF hyper expressed hMSCs animals showed: 1. An enhanced functional recovery	Horita et al., 2006
6	HGF (gene modification)	rMSCs	HGF hyper expressed rMSCs animals showed: 1. Improvement in neurological score 2. Reduced infarct volume 3. Decreased apoptosis-positive cells 4. Increased in cortex neuronal number	Zhao et al., 2006
7	Angiopoietin-1 (gene modification)	hMSCs	Angiopoietin-1 expressing hMSCs showed receiving rodents showed: 1. Enhanced angiogenesis 2. Improved functional outcome	Onda et al., 2008
8	FGF1 (gene modification)	rMSCs	FGF1 over expressing rMSCs receiving rodents showed: 1. Improved neurological function 2. Increased density of FGF1 protein 3. Decreased infarct volume 4. Decreased apoptotic index	Ghazavi et al., 2017
9	PlGF (gene modification)	hMSCs	Rodents receiving PlGF over expressed hMSCs showed: 1. Enhanced angiogenesis 2. Increased functional improvement	Liu et al., 2006

S no	Source of priming	Cell type	Effect	References
10	BDNF and VEGF (gene modification)	rMSCs	Animals receiving BDNF and VEGF co-expressing hMSCs showed: 1. Reduction in brain pathology 2. Enhanced neuroal functional performance	
11	VEGF (gene modification)	rMSCs	Animals receiving VEGF over expressed hMSCs showed: 1. Improved Functional recovery 2. Lower infarct volume	Miki et al., 2007
12	BDNF (gene modification)	hMSCs	Rodents receiving BDNF over expressing hMSCs showed: 1. Enhanced proliferation of endogenous neural stem cells 2. Enhanced differentiation of doublecortin (DCX-) positive neuroblasts 3. High number of Neuronal (NeuN-) positive mature cells 4. Greater functional recovery	Jeong et al., 2014
13	BDNF (gene modification)	hMSCs	Animals receiving BDNF over expressing hMSCs showed: 1. Reduced volume 2. Elicited functional improvement	Nomura et al., 2005
14	Stroke serum	rMSCs	Stroke serum primed rMSCs exhibited: 1. Significantly increased expression of cell proliferation associated miR-20a.	Kim et al., 2016
15	Stroke serum	hMSCs	Stroke serum primed hMSCs receiving rats exhibited: 1. VEGF, GDNF, and FGH higher expression 2. Increased neurogenesis 3. Increased angiogenesis 4. Enhanced behavioral improvements	Moon et al., 2018
16	Autologous stroke serum	hMSCs	Autologous stroke serum preconditioned hMSC receiving patients showed: 1. >20% reduced mean lesion volume after 1-week post-cell infusion	Honmou et al., 2011
17	CXCR4 (gene modification)	rMSCs	Rats receiving CXCR4 over-expressing rMSCs showed: 1. Increased capillary vascular volume 2. Improved neurological function	Yu et al., 2012

Table 2. (Continued)

S no	Source of priming	Cell type	Effect	References
18	CCR2 (gene modification)	hMSCs	Animals receiving CCR2 over-expressing hMSCs demonstrated: 1. Enhanced migration of CCR2-MSC cells towards the ischemic lesions 2. Improved neurological outcomes 3. Reduced Brain edema and blood-brain barrier (BBB) leakage levels	Huang et al., 2018

For instance, Redondo-Castro et al., (2017) showed that five minutes of priming with IL-1α led the MSC secretome to acquire a more anti-inflammatory phenotype that directed the inflamed mouse microglia to secrete reduced TNF-α and IL-6. Likewise, the anti-inflammatory interleukin over-expressing MSCs have been used in ischemic settings to investigate the beneficial effects. The preclinical stroke models have proven that IL-10 is an anti-inflammatory cytokine, which can induce immune tolerance, thereby, possessing the neuroprotective ability, through decreasing the pro-inflammatory signaling post cerebral ischemia (Liesz et al., 2014; Ooboshi et al., 2005). Neuroprotective administration of IL-10-MSCs has also resulted in reduced neuronal degeneration and increased functional regeneration.

In stroke recovery, the homing of MSC to brain infarcted sites is far from ideal (Chen et al., 2001b; Parekkadan et al., 2010). Stromal cell-derived factor-1 (SDF-1), secreted by astrocytes and endothelial cells around the infarcted region, attracts stem cells (Imitola et al., 2004). CXC chemokine receptor 4 (CXCR4) is a chemokine receptor specific for SDF1; the pretreatment of MSCs with SDF1 enhances MSC homing into the infarcted area (Wang et al., 2008).

MSC AND HYPOXIA PRE-TREATMENT

Exposure to sub-lethal hypoxic events can enhance the confrontation of tissues, organs, and also organisms from succeeding lethal injury caused by the hypoxia. This phenomenon is termed hypoxic or ischemic preconditioning and is well recognized in the cardium and the brain. Hypoxia preconditioning of stem cells before their use in therapy is an adaptive way that treats them to survive in the harsh post-ischemic local immune responses. The term hypoxia, regarding cell culture, refers to oxygen tensions ranging from 0 to 10% (Ejtehadifar et al., 2015). The physiological oxygen tension in tissues can differ from 1% in cartilage-bone marrow to 12% in peripheral blood (Das et al., 2010). Hence, the 21% O_2 routinely used for MSCs culture is far from the oxygen percentage received by these cells at their physiological niche. Therefore, employing the above cultural conditions to empower MSCs enhances their capability to endure for more extended periods, enhances their proliferation speed, and maintains them in an undifferentiated state. In general, it enhances MSCs' regenerative and cytoprotective effects (Hu et al., 2016; Lan et al., 2015). For instance, hypoxia preconditioning at 2.5 percent O_2 for 15 mins followed by oxygenation at 21 percent O_2 for half an hour, pursued by reoxygenation preconditioning at 2.5 percent O_2 for 72 hours, greatly enhances MSC proliferation and migration, *in vitro* (Kheirandish et al., 2017; Berniakovich and Giorgio, 2013). Nevertheless, hypoxic preconditioning does not interfere with the multipotent potential of MSCs, thereby, improving the ability of MSCs to survive in the oxidative and inflammatory microenvironment found within the stroke brain sites upon transplantation (Hawkins et al., 2013). The advantageous effect of hypoxic preconditioning is attributed to the fact that the MSCs respond to the hypoxic microenvironment through the upregulation of the transcription factor HIF-1α (Kiani et al., 2013). Stabilization of HIF-1α appears to depend on the hypoxia-induced increase of phosphorylated Akt- and p38 mitogen-activated protein kinase (p38MAPK) (Das et al., 2010). Akt is well known to contribute to self-renewal and delineation in stem cells

(Rivera et al., 2016). A schematic diagram depicting the effect of hypoxia precondition of MSCs is given in Figure 3. Table 3 summarizes the employment of hypoxia preconditioned MSCs in ischemic stroke.

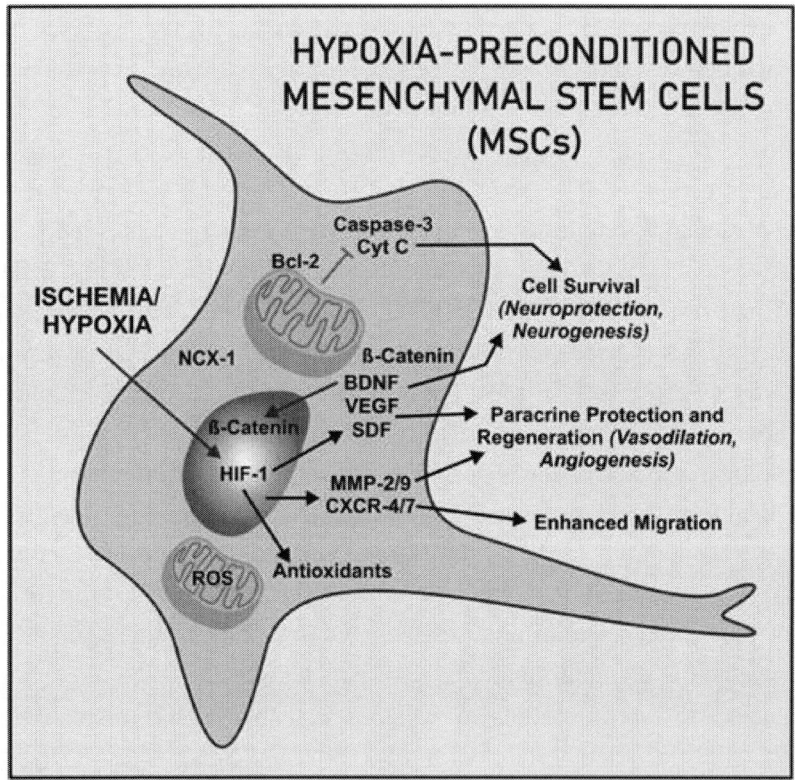

Figure 3. An ischemic representation of the mechanisms underlying the beneficial effects of hypoxic preconditioning in human MSCs. The hypoxic preconditioning approaches are designed to mimic and utilize endogenous protective tools to stimulate neuroprotection, cell- and tissue -regeneration, and brain function recovery. For instance, hypoxic preconditioning directly enhances HIF-1 upregulation that, in turn, induces BDNF, SDF-1, VEGF, and other genes, enhancing neurogenesis, angiogenesis, vasodilation, and increased cell survival. HIF-1 expression regulates antioxidants, survival signals, and other genes related to cell adhesion, polarization, migration, and anti-inflammatory responses. Partially adapted from a previous publication (Wei et al., 2017).

Table 3. Hypoxia preconditioning of MSCs

S no.	Source of priming	Cell type	Effect	Reference
1	Hypoxia	Bone marrow mesenchymal stem cells (BM-MSCs)	Hypoxia primed BM-MSCs receiving animals showed an enhanced: 1. Enhanced expression and release of trophic factors 2. Suppressed microglial activities 3. More matured Neurons 4. High rate of differentiation of BM-MSCs 5. Enhanced behavioural performance of BM-MSCs receiving rodents	Wei et al., 2012
2	Hypoxia	BM-MSCs	1. Hypoxia preconditioned BM-MSCs showed a reduced cell death number in the peri-infarct region 2. Hypoxia preconditioned BM-MSCs showed a reduced mice infarct volume 3. Hypoxia preconditioned BM-MSCs showed enhanced migratory capacities of BM-MSCs *in vitro* and *in vivo*	Wei et al., 2013
3	Hypoxia	BM-MSCs	1. Hypoxia preconditioned BM-MSCs promoted neural differentiation 2. Hypoxia preconditioning reduced the rate of apoptosis in BM-MSCs 3. Hypoxia preconditioning improved BM-MSCs proliferation and migration	Yu et al., 2016
4	Hypoxia	Bone marrow mesenchymal stem cells	1. Hypoxia preconditioned BM-MSCs enhanced endothelial cell proliferation, migration, and tube formation. 2. Hypoxia preconditioning BM-MSCs survived more and had more reparative capability	Theus et al., 2008
5	Hypoxia	BM-MSCs	1. Hypoxia preconditioned enhanced survival and differentiation of transplanted BMSCs 2. Hypoxia preconditioning enhanced the therapeutic efficacy of the BM-MSCs 3. Hypoxia preconditioned BM-MSCs showed a better neurological functional outcome in BM-MSCs group rodents	Chen et al., 2017
6	Hypoxia	BM-MSCs	1. Greater cell viability 2. Less pro-inflammatory cytokine secretion 3. Greater myelin production	Zarriello et al., 2019

Hypoxia preconditioning improves BM-MSCs targeted migration and survival. After intranasal delivery of hypoxia preconditioned BM-MSCs in a mouse focal cerebral stroke model, the expression of CXCR4, matrix metalloproteinase 2, and matrix metalloproteinase 9 increased, infarct volume in the peri-infarct region reduced, leading to neurological recovery (Wei et al., 2013). When preconditioned and non-preconditioned MSC was exposed to six hours of lethal anoxia, the number of preconditioned cells was found to be greater than that of the non-preconditioned cells. The high viability noticed was attributed to ischemic preconditioning induced activation of Akt/HIF-1α and the combined participation of hypoxia-responsive miR-107 and miR-210 mouse myocardial infarction model (Kim et al., 2012). Similarly, when hypoxic preconditioned human umbilical cord blood hematopoietic stem cells (hUCB) were transplanted in a stroke rodent model, hypoxia upregulated Epac1 through HIF-1α induction in hUCB. This improved therapeutic efficacy was correlated with improved engraftment and differentiation of hUCB in the receptive brain.

The transplant also improved cerebral blood flow into the ischemic brain via induction of angiogenesis (Lin et al., 2013). Likewise, bone marrow-derived mononuclear cells (MNCs) isolated from post-stroke rodents compared to the pre-stroke group significantly reduced the population of T- cells. In contrast, a significant increase in CD34[+] and natural killer cells were noticed. The concentrations of IL-10, IL-6, MCP-1, VEGF, and TNF-α were significantly high in post-stroke MNCs than pre-stroke MNCs. Compared to pre-stroke MNCs, post-stroke MNCs lead to a more significant recovery of neurological function and reduced lesion size (Yang et al., 2012). Similar results were obtained from other cell types employed in stroke recovery where ischemic or hypoxic preconditioning have endorsed tolerance and reparative properties of transplanted stem cells, therefore, raising their resistance to low substrates and oxygen accessibility in degenerating tissue via adaptive responses, such as the upregulation of anti-apoptotic genes (Bcl-2 and HIF-1) and reduction in caspase-3 activity (Wei et al., 2017). Therefore, in the light of the above findings, it is reasonable to predicate that hypoxia preconditioning empowers MSCs through activating their regenerative and reparative factors.

PRECONDITIONING OF MSCS WITH PHARMACOLOGICAL DRUGS AND CHEMICALS

The preconditioning of cells with drugs and specific chemicals has attracted the attention of the scientific community as a potential priming approach for *in vivo* experimentations. It is presumed that preconditioning with drugs is responsible for defending against ischemic damage during stem cell transplantation; hence, activating endogenous cell regeneration machinery. It has been demonstrated that certain drugs/chemicals exert protective effects on MSCs through the aforementioned approaches (Figure 4). Table 4 summarizes the use of pharmacological agents or drugs pre-treatment for MSCs activation prior to their use in ischemic stroke. However, the concentration of the pharmacological or chemical agents must be taken into the account as the function of MSCs may be impaired by extremely high concentrations of drugs. For example, high zoledronic acid concentrations inhibits the proliferation and osteogenic division of MSCs derived from bone marrow, while low zoledronic acid concentrations performs the opposite function without affecting their immune-modulatory properties (Hu et al., 2017). MSCs isolated from those with moderate but not high-risk myelodysplastic syndrome shows a lower formation of early hematopoietic progenitors in erythroid and myeloid colonies; however, preconditioning with lenalidomide successfully rescues dysfunction in MSCs derived from the disease patients of myelodyplastic syndrome (Ferrer et al., 2013).

Thus, the selection of an appropriate priming agent and proper optimization is a prerequisite before transplantation of 'Activated MSCs.' In this section, we will discuss a few attempts to use preconditioned MSCs with an aim to develop a potential therapeutic technique for stroke management.

Table 4. Preconditioning through pharmacological agents and chemicals

S. No.	Source of Priming	Cell type	Effects	Reference no.
1	Melatonin priming	rMSCs	1. Melatonin primed MSC showed a high survival rate 2. Melatonin primed MSC receiving group showed reduced brain infarction 3. Melatonin primed MSC receiving group showed enhanced angiogenesis 4. Melatonin primed MSC receiving group showed increased neurogenesis 5. Melatonin primed MSC receiving group showed increased vascular endothelial growth factor	Tang et al., 2014
2	CXCR4 gene modification	hMSCs	1. CXCR4 expressing hMSCs showed enhanced migration 2. Pronounced behavioral recovery was observed in rats those received CXCR4-hMSCs	Bang et al., 2012
3	C–C motif chemokine ligand 2 (CCL2)	hMSCs	CCL2-overexpressing MSCs receiving animals showed: 1. A smaller stroke volume 2. Enhanced functional recovery 3. Enhanced migration of cells into areas of higher CCR2 expression 4. Increased angiogenesis and endogenous neurogenesis	Lee et al., 2020
4	Neurogenin-1 (Ngn1)	hMSCs	Neurogenin-1 overexpressed hMSCs showed: 1. High homing potential 2. Enhanced cell-engraftment efficiency 3. Improved functional recovery in animals receiving neurogenin-1 overexpressed hMSCs	Kim et al., 2020

In the past few years, pharmacological agents, such as (e.g., lipopolysaccharide (LPS), NMDA, and the 3-nitropropionic acid (3-NPA, a metabolic inhibitor) have gained substantial experimental attention. Several studies have reported the beneficial effects of MSCs preconditioned with LPS and low doses of endotoxin against permanent and transient focal and neonatal hypoxia–ischemia stroke models of rodents (Bastide et al., 2003; Bordet et al. 2000; Eklind et al., 2005; Rosenzweig et al., 2007; Tasaki et al., 1997). Similar to the LPS, melatonin treated MSCs have also shown improvement in the animal model of stroke. In brief, Tang et al. investigated the effect of pretreatment of MSCs with melatonin in ischemic stroke rodents. Melatonin activated MSCs after transplantation into the ischemic rat brain resulted in increased survival of MSCs via reducing the apoptotic cell death. Moreover, melatonin treated MSCs showed an impediment in brain infarction and improved neurobehavioral outcomes. Further, melatonin pre-treatment enhanced the level of p-ERK1/2 in MSCs. U0126, an inhibitor of ERK phosphorylation, overturned the defensive effects of melatonin, suggesting that melatonin enhanced MSC viability and function through triggering the ERK1/2 signaling pathway (Tang et al., 2014).

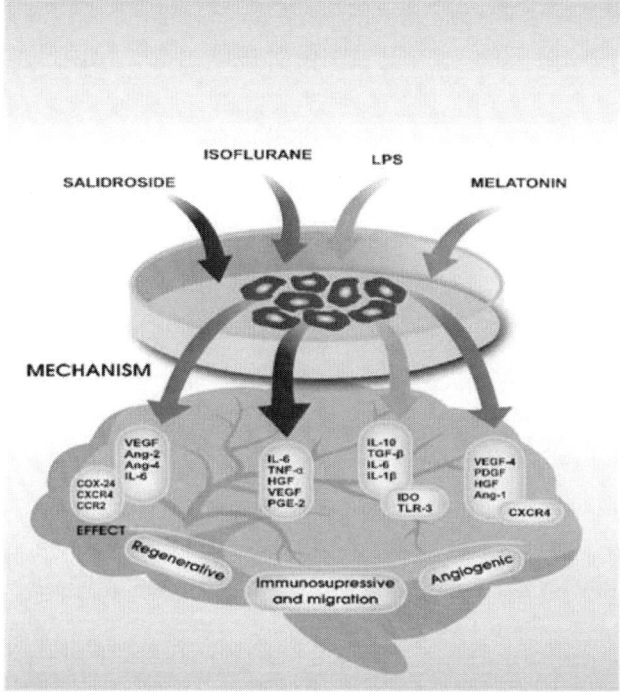

Figure 4. An overview of the production of 'Activated MSC' for the treatment of stroke. Preconditioning with pharmacological drugs and chemicals improves the therapeutic efficacy of MSCs. Preconditioning agents and the respectively triggered mechanisms are linked by arrows and boxes in the center of the brain clipart. The priming effects on MSC, i.e., regenerative, immunosuppressive, migratory, and angiogenesis, triggered by the pharmacological drugs and chemicals are indicated in the horizontal boxes at the lower interfaces of the water-shaded stroke brain (partially adapted from Noronha et al., 2019).

Tetramethylpyrazine (TMP), also known as ligustrazine and classified as an alkylpyrazine is a pharmacologically active component found in the rhizome of Chinese herb *Rhizoma Chuanxiong* (Chuanxiong) and fermented cocoa beans. The extracted TMP has been widely used as an anti-stroke agent in the clinical treatment of cerebrovascular and cardiovascular diseases in China [Hashim et al., 1998; Lin et al., 2015; Guo et al., 2014]. A derivative of TMP named Two-[[(1, 1-dimethylethyl) oxidoimino]-methyl]-3, 5, 6-trimethylpyrazine (TBN), possess significant activity of free radical scavenging in the cell-free chemical reaction system and MPP (+)-induced SH-SY5Y cell lines. TBN has shown penetration to BBB, which is a favorable property for stroke treatment (Sun et al., 2012; Guo et al., 2014; Zhang et al., 2016: Zhang et al., 2017). Attributing to interesting properties that can be exploited in stroke management, Li et al. (2017) used TMP as a preconditioning agent to prime MSCs and transplanted TMP activated MSCs in an MCAO model of the rat to investigate the beneficial effects. The transplanted MSCs, pretreated with TMP, exhibited induced migration and homing towards the infracted zone in comparison to non-preconditioned cells. In addition, increased expression of SDF-1 and CXCR4 was also observed by 'Activated MSCs' in the ischemic boundary zone, leading to improved neurological performance and enhanced angiogenesis. Like the above herbal treatment, Salidroside (p-hydroxyphenethyl-β-D-glucoside, Sal) a phenylpropanoid glycoside isolated from the *Rhodiola rosea*, commonly used as a conventional herbal remedy due to its strong anti-inflammatory and anti-apoptotic functions (Bai et al., 2014), was instrumental while MSCs where pre-treated with them for ischemic injury recovery. It was further proved that salidroside ameliorates brain ischemic injury by reducing neuroinflammation and neural damage (Zhang et al., 2019; Zhong et al., 2018; Zuo et al., 2018). It has been demonstrated that salidroside has a property to promote the proliferation of MSCs (Zhao et al., 2014a) and stimulate differentiation of MSCs into cholinergic neurons and dopaminergic neurons (Zhang et al., 2012; Zhao et al., 2014b). Evidence suggested that salidroside preconditioning, under hypoxia-ischemia, facilitated replication, migration, and inhibited the apoptosis of MSCs. In vivo studies showed that salidroside-pretreated MSCs improved therapeutic efficiency by improving neurogenesis and inhibiting neuroinflammation after ischemia in the hippocampal CA1 area, which was more noticeable compared to non-conditioned MSCs (Zhou et al., 2020).

The mood stabilizers valproate (VPA) and lithium have been used to treat bipolar disorder (Koch-Weser et al., 1980; Price et al., 1994). VPA inhibits histone deacetylases, which play a prominent role in transcriptional regulation, (Chuang et al., 2009) and lithium activates the Wnt downstream signaling pathway by inhibiting glycogen synthase kinase-3 (Chiu et al., 2010). Tsai and colleagues recently demonstrated that preconditioning of MSCs with VPA or lithium enhanced MSC migration by increasing CXCR4 expression via histone deacetylase inhibition or by elevating MMP-9 levels through glycogen synthase kinase-3 inhibition, respectively (Tsai et al., 2010). Likewise, isoflurane, an effective volatile anesthetic agent widely used in clinical practice when employed for MSCs in ischemic brain recovery, the MSCs pretreated with isoflurane showed effective migration towards the infarct region. To brief, Sun et al., (2015) reported that preconditioning of MSC with isoflurane (2% for 4 h) significantly improved the cell viability and migration potential. At the cellular level isoflurane priming unregulated the expression of HIF-1α and SDF-1 receptor CXCR4 in *in-vitro* and mediated the activation of Akt. When these cells were transplanted to the MCAO mouse model, they exhibited increased engraftment into the ischemic brain as well as improved functional recovery.

CONCLUSION

Although MSCs have an intrinsic ability to trigger or support the regeneration process in the damaged tissues, various extrinsic factors such as originating tissue, age and health condition of the donor, the batch of serum used during *in vitro* expansion of the MSCs, the number of passages, the concentration of oxygen, and the presence and absence of a pro-inflammatory niche at the time of cell administration are known to affect this capacity significantly (Hagmann et al., 2013; Hu et al., 2016). The preconditioning technique is based on the principle to train or prepare the cells to combat future unfavorable conditions. Thus *in vitro* priming of MSCs with numerous agents has been employed to improve the therapeutic ability of MSCs, which includes 3D culture, pharmacological compounds, inflammatory cytokines, and hypoxia. Even though preconditioning is probably the primary mechanism of MSC empowerment and functioning and may ultimately be proved as a unique therapeutic tool, the number of studies investigating the effect of preconditioning on MSCs is still deficient. Thus, for translation of preconditioning techniques from the lab to the clinics, future studies demand more detailed analyses to enhance the therapeutic potential of MSC post-transplantation.

REFERENCES

Aktas, E., Chamberlain, C. S., Saether, E. E., Duenwald-Kuehl, S. E., Kondratko-Mittnacht, J., Stitgen, M., Lee, J. S., Clements, A. E., Murphy, W. L., and Vanderby, R. 2017. "Immune Modulation with Primed Mesenchymal Stem Cells Delivered via Biodegradable Scaffold to Repair an Achilles Tendon Segmental Defect." *Journal of Orthopaedic Research* 35(2):269–280. doi: 10.1002/jor.23258.

Bai, H., Wang, C. B., Ma, X. H., Wei, Y. P., Xi, R., Zhao, Q., and Zhang, Q. 2014. "Effects of Salidroside on Proliferation of Bone Marrow Mesenchymal Stem Cells." *Zhongguo shi yan xue ye xue za zhi* 22(4):1072–1077. doi: 10.7534/j.issn.1009-2137.2014.04.035.

Bang, O. Y., Jin, K. S., Hwang, M. N., Kang, H. Y., Kim, B. J., Lee, S. J., Kang, S., Hwang, Y. K., Ahn, J. S., and Sung, K. W. 2012. "The Effect of CXCR4 Overexpression on Mesenchymal Stem Cell Transplantation in Ischemic Stroke." *Cell medicine* 4(2):65–76. doi: 10.3727/215517912X647172.

Bang, O. Y., Lee, J. S., Lee, P. H., and Lee, G. 2005. "Autologous Mesenchymal Stem Cell Transplantation in Stroke Patients." *Annals of Neurology* 57(6):874–882. doi: 10.1002/ana.20501.

Barone, F. C., and Feuerstein, G. Z. 1999. "Inflammatory Mediators and Stroke: New Opportunities for Novel Therapeutics." *Journal of Cerebral Blood Flow and Metabolism* 19(8):819–834. doi: 10.1097/00004647-199908000-00001.

Barrachina, L., Remacha, A. R., Romero, A., Vázquez, F. J., Albareda, J., Prades, M., Gosálvez, J., Roy, R., Zaragoza, P., Martín-Burriel, I., and Rodellar, C. 2017. "Priming Equine Bone Marrow-Derived Mesenchymal Stem Cells with Proinflammatory Cytokines: Implications in Immunomodulation-Immunogenicity Balance, Cell Viability, and Differentiation Potential." *Stem Cells and Development* 26(1):15–24. doi: 10.1089/scd.2016.0209.

Bastide, M., Gelé, P., Pétrault, O., Pu, Q., Caliez, A., Robin, E., Deplanque, D., Duriez, P., and Bordet, R. 2003. "Delayed Cerebrovascular Protective Effect of Lipopolysaccharide in Parallel to Brain Ischemic Tolerance." *Journal of Cerebral Blood Flow and Metabolism* 23(4):399–405. doi: 10.1097/01.WCB.0000050064.57184.F2.

Beegle, J., Lakatos, K., Kalomoiris, S., Stewart, H., Isseroff, R. R., Nolta, J. A., and Fierro, F. A. 2015. "Hypoxic Preconditioning of Mesenchymal Stromal Cells Induces Metabolic Changes, Enhances Survival, and Promotes Cell Retention in vivo." *Stem Cells (Dayton, Ohio)* 33(6):1818–1828. doi: 10.1002/stem.1976.

Berniakovich, I., and Giorgio, M. 2013. "Low Oxygen Tension Maintains Multipotency, whereas Normoxia Increases Differentiation of Mouse Bone Marrow Stromal Cells." *International Journal of Molecular Sciences* 14(1):2119–2134. doi: 10.3390/ijms14012119.

Bhasin, A., Kumaran, S. S., Bhatia, R., Mohanty, S., and Srivastava, M. 2017. "Safety and Feasibility of Autologous Mesenchymal Stem Cell Transplantation in Chronic Stroke in Indian Patients. A four-year follow up." *Journal of Stem Cells and Regenerative Medicine* 13(1):14–19. doi: 10.46582/jsrm.1301003.

Blakeley, J. O. and Llinas, R. H. 2007. "Thrombolytic Therapy for Acute Ischemic Stroke." *Journal of the Neurological Sciences* 261(1-2):55–62. doi: 10.1016/j.jns.2007.04.031.

Bordet, R., Deplanque, D., Maboudou, P., Puisieux, F., Pu, Q., Robin, E., Martin, A., Bastide, M., Leys, D., Lhermitte, M., and Dupuis, B. 2000. "Increase in Endogenous Brain Superoxide Dismutase as a Potential Mechanism of Lipopolysaccharide-Induced Brain Ischemic Tolerance." *Journal of Cerebral Blood Flow and Metabolism* 20(8):1190–1196. doi: 10.1097/00004647-200008000-00004.

Carvalho, A., Sousa, M., Alencar-Silva, T., Carvalho, J. L., and Saldanha-Araujo, F. 2019. "Mesenchymal Stem Cells Immunomodulation: The Road to IFN-γ Licensing and the Path Ahead." *Cytokine and Growth Factor Reviews* 47:32–42. doi: 10.1016/j.cytogfr.2019.05.006.

Chen, J., Li, Y., Wang, L., Zhang, Z., Lu, D., Lu, M., and Chopp, M. 2001b. "Therapeutic Benefit of Intravenous Administration of Bone Marrow Stromal Cells after Cerebral Ischemia in Rats." *Stroke* 32(4):1005–1011. doi: 10.1161/01.str.32.4.1005.

Chen, J., Sanberg, P. R., Li, Y., Wang, L., Lu, M., Willing, A. E., Sanchez-Ramos, J., and Chopp, M. 2001a. "Intravenous Administration of Human Umbilical Cord Blood Reduces Behavioral Deficits after Stroke in Rats." *Stroke* 32(11):2682–2688. doi: 10.1161/hs1101.098367.

Chen, J., Yang, Y., Shen, L., Ding, W., Chen, X., Wu, E., Cai, K., and Wang, G. 2017. "Hypoxic Preconditioning Augments the Therapeutic Efficacy of Bone Marrow Stromal Cells in a Rat Ischemic Stroke Model." *Cellular and Molecular Neurobiology* 37(6):1115–1129. doi: 10.1007/s10571-016-0445-1.

Chiu, C. T., and Chuang, D. M. 2010. "Molecular Actions and Therapeutic Potential of Lithium in Preclinical and Clinical Studies of CNS Disorders." *Pharmacology and Therapeutics* 128(2):281–304. doi: 10.1016/j.pharmthera.2010.07.006.

Chuang, D. M., Leng, Y., Marinova, Z., Kim, H. J., and Chiu, C. T. 2009. "Multiple Roles of HDAC Inhibition in Neurodegenerative Conditions.

Trends in Neurosciences 32(11):591–601. doi: 10.1016/j.tins.2009.06. 002.

Courties, G., Herisson, F., Sager, H. B., Heidt, T., Ye, Y., Wei, Y., Sun, Y., Severe, N., Dutta, P., Scharff, J., Scadden, D. T., Weissleder, R., Swirski, F. K., Moskowitz, M. A., and Nahrendorf, M. 2015. "Ischemic Stroke Activates Hematopoietic Bone Marrow Stem Cells." *Circulation Research* 116(3):407–417. doi: 10.1161/CIRCRESAHA. 116.305207.

Cunningham, C. J., Redondo-Castro, E., and Allan, S. M. 2018. "The Therapeutic Potential of the Mesenchymal Stem Cell Secretome in Ischaemic Stroke." *Journal of Cerebral Blood Flow and Metabolism* 38(8):1276–1292. doi: 10.1177/0271678X18776802.

Cunningham, C. J., Wong, R., Barrington, J., Tamburrano, S., Pinteaux, E., and Allan, S. M. 2020. "Systemic Conditioned Medium Treatment from Interleukin-1 Primed Mesenchymal Stem Cells Promotes Recovery after Stroke." *Stem Cell Research and Therapy* 11(1):32. doi: 10.1186/s13287-020-1560-y.

da Silva Meirelles, L., Chagastelles, P. C., and Nardi, N. B. 2006. "Mesenchymal Stem Cells Reside in Virtually All Post-natal Organs and Tissues." *Journal of Cell Science* 119(Pt 11), 2204–2213. doi: 10.1242/jcs.02932.

Das, R., Jahr, H., van Osch, G. J., and Farrell, E. 2010. "The Role of Hypoxia in Bone Marrow-derived Mesenchymal Stem Cells: Considerations for Regenerative Medicine Approaches." *Tissue Engineering Part B, Reviews* 16(2):159–168. doi: 10.1089/ten.TEB. 2009.0296.

Deng, L., Peng, Q., Wang, H., Pan, J., Zhou, Y., Pan, K., Li, J., Wu, Y., and Wang, Y. 2019. "Intrathecal Injection of Allogenic Bone Marrow-Derived Mesenchymal Stromal Cells in Treatment of Patients with Severe Ischemic Stroke: Study Protocol for a Randomized Controlled Observer-Blinded Trial." *Translational Stroke Research* 10(2):170–177. doi: 10.1007/s12975-018-0634-y.

Dominici, M., Le Blanc, K., Mueller, I., Slaper-Cortenbach, I., Marini, F., Krause, D., Deans, R., Keating, A., Prockop, D. j., and Horwitz, E.

2006. "Minimal Criteria for Defining Multipotent Mesenchymal Stromal Cells. The International Society for Cellular Therapy Position Statement." *Cytotherapy* 8(4):315–317. doi: 10.1080/146532406 00855905.

Du, L., Yu, Y., Ma, H., Lu, X., Ma, L., Jin, Y., and Zhang, H. 2014. "Hypoxia Enhances Protective Effect of Placental-derived Mesenchymal Stem Cells on Damaged Intestinal Epithelial Cells by Promoting Secretion of Insulin-like Growth Factor-1." *International Journal of Molecular Sciences* 15(2):1983–2002. doi: 10.3390/ijms 15021983.

Dulamea A. O. 2015. "The Potential use of Mesenchymal Stem Cells in Stroke Therapy–From Bench to Bedside." *Journal of the Neurological Sciences* 352(1-2):1–11. doi: 10.1016/j.jns.2015.03.014.

Ejtehadifar, M., Shamsasenjan, K., Movassaghpour, A., Akbarzadehlaleh, P., Dehdilani, N., Abbasi, P., Molaeipour, Z., and Saleh, M. 2015. "The Effect of Hypoxia on Mesenchymal Stem Cell Biology." *Advanced Pharmaceutical Bulletin* 5(2):141–149. doi: 10.15171/apb. 2015.021.

Eklind, S., Mallard, C., Arvidsson, P., and Hagberg, H. 2005. "Lipopolysaccharide Induces Both a Primary and a Secondary Phase of Sensitization in the Developing Rat Brain." *Pediatric Research* 58(1):112–116. doi: 10.1203/01.PDR.0000163513.03619.8D.

Ferrer, R. A., Wobus, M., List, C., Wehner, R., Schönefeldt, C., Brocard, B., Mohr, B., Rauner, M., Schmitz, M., Stiehler, M., Ehninger, G., Hofbauer, L. C., Bornhäuser, M., and Platzbecker, U. 2013. "Mesenchymal Stromal Cells from Patients with Myelodyplastic Syndrome Display Distinct Functional Alterations That are Modulated by Lenalidomide." *Haematologica* 98(11):1677–1685. doi: 10.3324/ haematol.2013.083972.

Friis, T., Haack-Sørensen, M., Mathiasen, A. B., Ripa, R. S., Kristoffersen, U. S., Jørgensen, E., Hansen, L., Bindslev, L., Kjær, A., Hesse, B., Dickmeiss, E., and Kastrup, J. 2011. "Mesenchymal Stromal Cell Derived Endothelial Progenitor Treatment in Patients with Refractory

Angina." *Scandinavian Cardiovascular Journal: SCJ* 45(3):161–168. doi: 10.3109/14017431.2011.569571.

Ghazavi, H., Hoseini, S. J., Ebrahimzadeh-Bideskan, A., Mashkani, B., Mehri, S., Ghorbani, A., Sadri, K., Mahdipour, E., Ghasemi, F., Forouzanfar, F., Hoseini, A., Pasdar, A. R., Sadeghnia, H. R., and Ghayour-Mobarhan, M. 2017. "Fibroblast Growth Factor Type 1 (FGF1)-Overexpressed Adipose-Derived Mesenchaymal Stem Cells (AD-MSC^{FGF1}) Induce Neuroprotection and Functional Recovery in a Rat Stroke Model." *Stem Cell Reviews and Reports* 13(5):670–685. doi: 10.1007/s12015-017-9755-z.

Gronthos, S., Brahim, J., Li, W., Fisher, L. W., Cherman, N., Boyde, A., DenBesten, P., Robey, P. G., and Shi, S. 2002. "Stem Cell Properties of Human Dental Pulp Stem Cells." *Journal of Dental Research* 81(8):531–535. doi: 10.1177/154405910208100806.

Guo, B., Xu, D., Duan, H., Du, J., Zhang, Z., Lee, S. M., and Wang, Y. 2014. "Therapeutic Effects of Multifunctional Tetramethylpyrazine Nitrone on Models of Parkinson's Disease in vitro and in vivo." *Biological and Pharmaceutical Bulletin* 37(2):274–285. doi: 10.1248/bpb.b13-00743.

Hagmann, S., Moradi, B., Frank, S., Dreher, T., Kämmerer, P. W., Richter, W., and Gotterbarm, T. 2013. "Different Culture Media Affect Growth Characteristics, Surface Marker Distribution and Chondrogenic Differentiation of Human Bone Marrow-derived Mesenchymal Stromal Cells." *BMC Musculoskeletal Disorders* 14:223. doi: 10.1186/1471-2474-14-223.

Hashim, P., Jinap, S., Sharifah, K. S. M., and Asbi, A. 1998. "Effect of Mass and Turning Time on Free Amino Acid, Peptide-N, Sugar and Pyrazine Concentration during Cocoa Fermentation." *Journal of the Science of Food and Agriculture* 78:543–550. doi: 10.1002/(SICI)1097-0010(199812)78:4<543::AID-JSFA152>3.0.CO;2-2.

Hawkins, K. E., Sharp, T. V., and McKay, T. R. 2013. "The Role of Hypoxia in Stem Cell Potency and Differentiation." *Regenerative Medicine* 8(6):771–782. doi: 10.2217/rme.13.71.

Honmou, O., Houkin, K., Matsunaga, T., Niitsu, Y., Ishiai, S., Onodera, R., Waxman, S. G., and Kocsis, J. D. 2011." Intravenous Administration of Auto Serum-Expanded Autologous Mesenchymal Stem Cells in Stroke." *Brain: a Journal of Neurology* 134(Pt 6):1790–1807. doi: 10.1093/brain/awr063.

Horie, N., Pereira, M. P., Niizuma, K., Sun, G., Keren-Gill, H., Encarnacion, A., Shamloo, M., Hamilton, S. A., Jiang, K., Huhn, S., Palmer, T. D., Bliss, T. M., and Steinberg, G. K. 2011. "Transplanted Stem Cell-Secreted Vascular Endothelial Growth Factor Effects Poststroke Recovery, Inflammation, and Vascular Repair." *Stem Cells (Dayton, Ohio)* 29(2):274–285. doi: 10.1002/stem.584.

Horita, Y., Honmou, O., Harada, K., Houkin, K., Hamada, H., and Kocsis, J. D. 2006. "Intravenous Administration of Glial Cell Line-Derived Neurotrophic Factor Gene-Modified Human Mesenchymal Stem Cells Protects Against Injury in a Cerebral Ischemia Model in the Adult Rat." *Journal of Neuroscience Research* 84(7):1495–1504. doi: 10.1002/jnr.21056.

Hu, C., and Li, L. 2018. "Preconditioning Influences Mesenchymal Stem Cell Properties in vitro and in vivo." *Journal of Cellular and Molecular Medicine* 22(3):1428–1442. doi:10.1111/jcmm.13492.

Hu, L., Wen, Y., Xu, J., Wu, T., Zhang, C., Wang, J., Du, J., and Wang, S. 2017. "Pretreatment with Bisphosphonate Enhances Osteogenesis of Bone Marrow Mesenchymal Stem Cells." *Stem Cells and Development* 26(2):123–132. doi:10.1089/scd.2016.0173.

Hu, X., Xu, Y., Zhong, Z., Wu, Y., Zhao, J., Wang, Y., Cheng, H., Kong, M., Zhang, F., Chen, Q., Sun, J., Li, Q., Jin, J., Li, Q., Chen, L., Wang, C., Zhan, H., Fan, Y., Yang, Q., Yu, L., and Wang, J. 2016. "A Large-Scale Investigation of Hypoxia-Preconditioned Allogeneic Mesenchymal Stem Cells for Myocardial Repair in Nonhuman Primates: Paracrine Activity without Remuscularization." *Circulation Research* 118(6):970–983. doi: 10.1161/CIRCRESAHA.115.307516.

Huang, C., Dai, J., and Zhang, X. A. 2015. "Environmental Physical Cues Determine the Lineage Specification of Mesenchymal Stem Cells."

Biochimica et Biophysica acta 1850(6):1261–1266. doi: 10.1016/j.bbagen.2015.02.011.

Huang, Y., Wang, J., Cai, J., Qiu, Y., Zheng, H., Lai, X., Sui, X., Wang, Y., Lu, Q., Zhang, Y., Yuan, M., Gong, J., Cai, W., Liu, X., Shan, Y., Deng, Z., Shi, Y., Shu, Y., Zhang, L., Qiu, W., and Xiang, A. P. 2018. "Targeted Homing of CCR2-Overexpressing Mesenchymal Stromal Cells to Ischemic Brain Enhances Post-Stroke Recovery Partially Through PRDX4-Mediated Blood-Brain Barrier Preservation." *Theranostics* 8(21):5929–5944. doi: 10.7150/thno.28029.

Imitola, J., Raddassi, K., Park, K. I., Mueller, F. J., Nieto, M., Teng, Y. D., Frenkel, D., Li, J., Sidman, R. L., Walsh, C. A., Snyder, E. Y., and Khoury, S. J. 2004. "Directed Migration of Neural Stem Cells to Sites of CNS Injury by the Stromal Cell-Derived Factor 1alpha/CXC Chemokine Receptor 4 Pathway." *Proceedings of the National Academy of Sciences of the United States of America* 101(52):18117–18122. doi: 10.1073/pnas.0408258102.

in 't Anker, P. S., Noort, W. A., Scherjon, S. A., Kleijburg-van der Keur, C., Kruisselbrink, A. B., van Bezooijen, R. L., Beekhuizen, W., Willemze, R., Kanhai, H. H., and Fibbe, W. E. 2003. "Mesenchymal Stem Cells in Human Second-Trimester Bone Marrow, Liver, Lung, and Spleen Exhibit a Similar Immunophenotype but a Heterogeneous Multilineage Differentiation Potential." *Haematologica* 88(8):845–852. https://pubmed.ncbi.nlm.nih.gov/12935972/.

Jaillard, A., Hommel, M., Moisan, A., Zeffiro, T. A., Favre-Wiki, I. M., Barbieux-Guillot, M., Vadot, W., Marcel, S., Lamalle, L., Grand, S., Detante, O., and (for the ISIS-HERMES Study Group) 2020. "Autologous Mesenchymal Stem Cells Improve Motor Recovery in Subacute Ischemic Stroke: a Randomized Clinical Trial." *Translational Stroke Research* 11(5):910–923. doi: 10.1007/s12975-020-00787-z.

Jeong, C. H., Kim, S. M., Lim, J. Y., Ryu, C. H., Jun, J. A., and Jeun, S. S. 2014. "Mesenchymal Stem Cells Expressing Brain-Derived Neurotrophic Factor Enhance Endogenous Neurogenesis in an

Ischemic Stroke Model." *BioMed Research International* 2014: 129145. doi: 10.1155/2014/129145.

Jiang, R., Han, Z., Zhuo, G., Qu, X., Li, X., Wang, X., Shao, Y., Yang, S., and Han, Z. C. 2011. "Transplantation of Placenta-Derived Mesenchymal Stem Cells in Type 2 Diabetes: a Pilot Study." *Frontiers of Medicine* 5(1):94–100. doi: 10.1007/s11684-011-0116-z.

Jiang, Y., Vaessen, B., Lenvik, T., Blackstad, M., Reyes, M., and Verfaillie, C. M. 2002. "Multipotent Progenitor Cells Can be Isolated from Postnatal Murine Bone Marrow, Muscle, and Brain." *Experimental Hematology* 30(8):896–904. doi: 10.1016/s0301-472x (02)00869-x.

Jin, K., Sun, Y., Xie, L., Mao, X. O., Childs, J., Peel, A., Logvinova, A., Banwait, S., and Greenberg, D. A. 2005."Comparison of Ischemia-Directed Migration of Neural Precursor Cells after Intrastriatal, Intraventricular, or Intravenous Transplantation in the Rat." *Neurobiology of Disease* 18(2):366–374. doi: 10.1016/j.nbd.2004. 10.010.

Kasahara, Y., Yamahara, K., Soma, T., Stern, D. M., Nakagomi, T., Matsuyama, T., and Taguchi, A. 2016. "Transplantation of Hematopoietic Stem Cells: Intra-arterial Versus Intravenous Administration Impacts Stroke Outcomes in a Murine Model. *Translational Research* 176:69–80. doi: 10.1016/j.trsl.2016.04.003.

Kheirandish, M., Gavgani, S. P., and Samiee, S. 2017. "The Effect of Hypoxia Preconditioning on the Neural and Stemness Genes Expression Profiling in Human Umbilical Cord Blood Mesenchymal Stem Cells." *Transfusion and Apheresis Science* 56(3):392–399. Doi: 10.1016/j.transci.2017.03.015.

Kiani, A. A., Kazemi, A., Halabian, R., Mohammadipour, M., Jahanian-Najafabadi, A., and Roudkenar, M. H. 2013. "HIF-1α Confers Resistance to Induced Stress in Bone Marrow-Derived Mesenchymal Stem Cells." *Archives of Medical Research* 44(3):185–193. doi: 10.1016/j.arcmed.2013.03.006.

Kim, E. H., Kim, D. H., Kim, H. R., Kim, S. Y., Kim, H. H., and Bang, O. Y. 2016. "Stroke Serum Priming Modulates Characteristics of

Mesenchymal Stromal Cells by Controlling the Expression miRNA-20a." *Cell Transplantation* 25(8):1489–1499. doi: 10.3727/09636891 6X690430.

Kim, G. H., Subash, M., Yoon, J. S., Jo, D., Han, J., Hong, J. M., Kim, S. S., and Suh-Kim, H. (2020). "Neurogenin-1 Overexpression Increases the Therapeutic Effects of Mesenchymal Stem Cells through Enhanced Engraftment in an Ischemic Rat Brain. *International Journal of Stem Cells* 13(1):127–141. doi: 10.15283/ijsc19111.

Kim, H. W., Mallick, F., Durrani, S., Ashraf, M., Jiang, S., and Haider, K. H. 2012. "Concomitant Activation of miR-107/PDCD10 and Hypoxemia-210/Casp8ap2 and Their Role in Cytoprotection During Ischemic Preconditioning of Stem Cells." *Antioxidants and redox signaling* 17(8):1053–1065. doi: 10.1089/ars.2012.4518.

Ko, K. I., Coimbra, L. S., Tian, C., Alblowi, J., Kayal, R. A., Einhorn, T. A., Gerstenfeld, L. C., Pignolo, R. J., and Graves, D. T. 2015. "Diabetes Reduces Mesenchymal Stem Cells in Fracture Healing Through a TNFα-Mediated Mechanism." *Diabetologia*, 58(3):633–642. doi: 10.1007/s00125-014-3470-y.

Koch-Weser, J., and Browne, T. R. 1980. Drug Therapy: Valproic Acid. *The New England Journal of Medicine* 302(12):661–666. doi: 10.1056/NEJM198003203021204.

Kurozumi, K., Nakamura, K., Tamiya, T., Kawano, Y., Kobune, M., Hirai, S., Uchida, H., Sasaki, K., Ito, Y., Kato, K., Honmou, O., Houkin, K., Date, I., and Hamada, H. 2004. "BDNF Gene-Modified Mesenchymal Stem Cells Promote Functional Recovery and Reduce Infarct Size in the Rat Middle Cerebral Artery Occlusion Model." *Molecular Therapy: the Journal of the American Society of Gene Therapy*, 9(2):189–197. doi: 10.1016/j.ymthe.2003.10.012.

Lan, Y. W., Choo, K. B., Chen, C. M., Hung, T. H., Chen, Y. B., Hsieh, C. H., Kuo, H. P., and Chong, K. Y. 2015. "Hypoxia-Preconditioned Mesenchymal Stem Cells Attenuate Bleomycin-Induced Pulmonary Fibrosis." *Stem Cell Research and Therapy* 6(1):97. doi: 10.1186/s13287-015-0081-6.

Lee, S. H., Jin, K. S., Bang, O. Y., Kim, B. J., Park, S. J., Lee, N. H., Yoo, K. H., Koo, H. H., and Sung, K. W. 2015. "Differential Migration of Mesenchymal Stem Cells to Ischemic Regions after Middle Cerebral Artery Occlusion in Rats." *PloS One*, 10(8):e0134920. doi: 10.1371/journal.pone.0134920.

Lee, S., Kim, O. J., Lee, K. O., Jung, H., Oh, S. H., and Kim, N. K. 2020. "Enhancing the Therapeutic Potential of *CCL2*-Overexpressing Mesenchymal Stem Cells in Acute Stroke." *International Journal of Molecular Sciences* 21(20):7795. doi: 10.3390/ijms21207795.

Lehrke, S., Mazhari, R., Durand, D. J., Zheng, M., Bedja, D., Zimmet, J. M., Schuleri, K. H., Chi, A. S., Gabrielson, K. L., and Hare, J. M. 2006. "Aging Impairs the Beneficial Effect of Granulocyte Colony-Stimulating Factor and Stem Cell Factor on Post-Myocardial Infarction Remodeling." *Circulation Research* 99(5):553–560. doi: 10.1161/01.RES.0000238375.88582.d8.

Li, L., Chu, L., Fang, Y., Yang, Y., Qu, T., Zhang, J., Yin, Y., and Gu, J. 2017. "Preconditioning of Bone Marrow-Derived Mesenchymal Stromal Cells by Tetramethylpyrazine Enhances Cell Migration and Improves Functional Recovery After Focal Cerebral Ischemia in Rats." *Stem Cell Research and Therapy* 8(1):112. doi: 10.1186/s13287-017-0565-7.

Liesz, A., Bauer, A., Hoheisel, J. D., and Veltkamp, R. 2014. "Intracerebral Interleukin-10 Injection Modulates Post-Ischemic Neuroinflammation: An Experimental Microarray Study." *Neuroscience Letters* 579:18–23. doi: 10.1016/j.neulet.2014.07.003.

Lin, C. H., Lee, H. T., Lee, S. D., Lee, W., Cho, C. W., Lin, S. Z., Wang, H. J., Okano, H., Su, C. Y., Yu, Y. L., Hsu, C. Y., and Shyu, W. C. 2013. Role of HIF-1α-Activated Epac1 on HSC-Mediated Neuroplasticity in a Stroke Model." *Neurobiology of Disease* 58:76–91. doi: 10.1016/j.nbd.2013.05.006.

Lin, J. B., Zheng, C. J., Zhang, X., Chen, J., Liao, W. J., and Wan, Q. 2015. "Effects of Tetramethylpyrazine on Functional Recovery and Neuronal Dendritic Plasticity after Experimental Stroke." *Evidence-*

based Complementary and Alternative Medicine 2015:394926. doi: 10.1155/2015/394926.

Liu, H., Honmou, O., Harada, K., Nakamura, K., Houkin, K., Hamada, H., and Kocsis, J. D. 2006. "Neuroprotection by PlGF Gene-Modified Human Mesenchymal Stem Cells after Cerebral Ischaemia." *Brain: A Journal of Neurology* 129(Pt 10):2734–2745. doi: 10.1093/brain/awl207.

Loganathan, A., Linley, J. E., Rajput, I., Hunter, M., Lodge, J. P., and Sandle, G. I. 2011. "Basolateral Potassium (IKCa) Channel Inhibition Prevents Increased Colonic Permeability Induced by Chemical Hypoxia." *American Journal of Physiology. Gastrointestinal and Liver Physiology* 300(1):G146–G153. doi: 10.1152/ajpgi.00472.2009.

Lu, Z., Chen, Y., Dunstan, C., Roohani-Esfahani, S., and Zreiqat, H. 2017. "Priming Adipose Stem Cells with Tumor Necrosis Factor-Alpha Preconditioning Potentiates Their Exosome Efficacy for Bone Regeneration." *Tissue engineering* 23(21-22):1212–1220. doi: 10.1089/ten.tea.2016.0548.

Luo, R., Lu, Y., Liu, J., Cheng, J., and Chen, Y. 2019. "Enhancement of the Efficacy of Mesenchymal Stem Cells in the Treatment of Ischemic Diseases." *Biomedicine and pharmacotherapy* 109:2022–2034. doi: 10.1016/j.biopha.2018.11.068.

Magne, B., Dedier, M., Nivet, M., Coulomb, B, Banzet, S., Lataillade, J. J., and Trouillas, M. 2020. "IL-1β-Primed Mesenchymal Stromal Cells Improve Epidermal Substitute Engraftment and Wound Healing via Matrix Metalloproteinases and Transforming Growth Factor-β1." *Journal of Investigative Dermatology* 140(3):688–698.e21. doi: 10.1016/j.jid.2019.07.721.

Marder, V, J., Jahan, R., Gruber, T., Goyal, A., and Arora, V. 2010. "Thrombolysis with Plasmin Implications for Stroke Treatment." *Stroke* 41:S45-9. doi: 10.1161/STROKEAHA.110.595157.

Mastri, M., Lin, H., and Lee, T. 2014. "Enhancing the Efficacy of Mesenchymal Stem Cell Therapy." *World Journal of Stem Cells*. 6:82–93. doi: 10.4252/wjsc.v6.i2.82.

Maumus, M., Manferdini, C., Toupet, K., Peyrafitte, J. A., Ferreira, R., Facchini, A., Gabusi, E., Bourin, P., Jorgensen, C., Lisignoli, G., and Noel, D. 2013. "Adipose Mesenchymal Stem Cells Protect Chondrocytes from Degeneration Associated with Osteoarthritis." *Stem Cell Research* 11:834–844. doi: 10.1016/j.scr.2013.05.008.

Miki, Y., Nonoguchi, N., Ikeda, N., Coffin, R. S., Kuroiwa, T., and Miyatake, S. 2007. "Vascular Endothelial Growth Factor Gene-Transferred Bone Marrow Stromal Cells Engineered with a Herpes Simplex Virus Type 1 Vector can Improve Neurological Deficits and Reduce Infarction Volume in Rat Brain Ischemia." *Neurosurgery* 61(3):586-94; discussion 594-5. doi: 10.1227/01.NEU.0000290907.30814.42.

Moon, G, J., Cho, Y. H., Kim, D. H., Sung, J. H., Son, J. P., Kim, S., Cha, J. M., and Bang, O. Y. 2018. "Serum-Mediated Activation of Bone Marrow-Derived Mesenchymal Stem Cells in Ischemic Stroke Patients A Novel Preconditioning Method." *Stem Cell Transplantation* 27(3):485-500. doi: 10.1177/0963689718755404.

Muir, K. W. 2017. "Clinical Trial Design for Stem Cell Therapies in Stroke: What Have We Learned?" *Neurochemistry International* 106:108–113. doi: 10.1016/j.neuin t.2016.09.011.

Musuka, T. D., Wilton, S. B., Traboulsi, M., and Hill, M. D. 2015. "Diagnosis and Management of Acute Ischemic Stroke Speed is Critical." *Canadian Medical Association Journal* 187(12):887-93. doi: 10.1503/cmaj.140355.

Nakajima, M., Nito, C., Sowa, K., Suda, S., Nishiyama, Y., Nakamura-Takahashi, A., Nitahara-Kasahara, Y., Imagawa, K., Hirato, T., Ueda, M., Kimura, K., and Okada, T. 2017. "Mesenchymal Stem Cells Overexpressing Interleukin-10 Promote Neuroprotection in Experimental Acute Ischemic Stroke." *Molecular Therapy Methods and Clinical Development* (6):102–111. doi: 10.1016/j.omtm.2017.06.005.

Nam, H. S., Kwon, I., Lee, B. H., Kim, H., Kim, J., An, S., Lee, O. H., Lee, P. H., Kim, H. O., Namgoong, H., Kim, Y. D., and Heo, J. H. 2015. "Effects of Mesenchymal Stem Cell Treatment on the

Expression of Matrix Metalloproteinases and Angiogenesis during Ischemic Stroke Recovery. *PLoS One* 10(12):e0144218. doi: 10.1371/journal. pone.0144218.

Nava, M. M., Raimondi, M. T., and Pietrabissa, R. 2012. "Controlling Self-Renewal and Differentiation of Stem Cells via Mechanical Cues." *Journal of Biomedicine and Biotechnology* 797410. doi: 10.1155/2012/797410.

Nichols, J. E., Niles, J. A., DeWitt, D., Prough, D., Parsley, M., Vega, S., Cantu, A., Lee, E., and Cortiella, J. 2013. "Neurogenic and Neuro-Protective Potential of a Novel Subpopulation of Peripheral Blood-Derived CD133+ ABCG2+CXCR4+ Mesenchymal Stem Cells: Development of Autologous Cell-Based Therapeutics for Traumatic Brain Injury." *Stem Cell Research and Therapy* 4(1):3. doi: 10.1186/scrt151.

Nomura, T., Honmou, O., Harada, K., Houkin, K., Hamada, H., and Kocsis, J. D. 2005. "Infusion of Brain-Derived Neurotrophic Factor Gene-Modified Human Mesenchymal Stem Cells Protects Against Injury in a Cerebral Ischemia Model in Adult Rat. *Neuroscience* 136(1):161-9. doi: 10.1016/j.neuroscience.2005.06.062.

Noronha, N. C., Mizukami, A., Caliári-Oliveira, C., Cominal, J. G., Rocha, J. L. M., Covas, D. T., Swiech, K., and Malmegrim, K. C. R. 2019. "Priming approaches to improve the efficacy of mesenchymal stromal cell-based therapies." *Stem Cell Research Therapy* 2;10(1):131. doi: 10.1186/s13287-019-1224-y.

Onda, T., Honmou, O., Harada, K., Houkin, K., Hamada, H., and Kocsis, J. D. 2008. "Therapeutic Benefits by Human Mesenchymal Stem Cells (hMSCs) and Ang-1 Gene-Modified hMSCs after Cerebral Ischemia." *Journal of Cerebral Blood Flow and Metabolism* 28(2):329-40. doi: 10.1038/sj.jcbfm.9600527.

Ooboshi, H., Ibayashi, S., Shichita, T., Kumai, Y., Takada, J., Ago, T., Arakawa, S., Sugimori, H., Kamouchi, M., Kitazono, T., and Iida, M. 2005. "Postischemic Gene Transfer of Interleukin-10 Protects Against Both Focal and Global Brain Ischemia." *Circulation* 111:913–919. doi: 10.1161/01.CIR.0000155622.68580. DC.

Oskowitz, A., McFerrin, H., Gutschow, M., Carter, M. L., and Pochampally, R. 2011. "Serum–deprived human multipotent mesenchymal stromal cells (MSCs) are highly angiogenic." *Stem Cell Research* 6:215–225. doi: 10.1016/j.scr.2011.01.004.

Pan, Q., Kuang, X., Cai, S., Wang, X., Du, D., Wang, J., Wang, Y., Chen, Y., Bihl, J., Chen, Y., Zhao, B., and Ma, X. 2020. "miR-132-3p Priming Enhances the Effects of Mesenchymal Stromal Cell-Derived Exosomes on Ameliorating Brain Ischemic Injury. *Stem Cell Research Therapy* 11(1):260. doi: 10.1186/s13287-020-01761-0.

Parekkadan, B., and Milwid, J. M. 2010. "Mesenchymal Stem Cells as Therapeutics." *Annual Review of Biomedical Engineering* 12:87–117. doi: 10.1146/annurev-bioeng-070909-105309.

Petrovic-Djergovic, D., Goonewardena, S. N., and Pinsky, D. J. 2016. "Inflammatory Disequilibrium in Stroke." *Circulation Research* 119(1):142-58. doi: 10.1161/CIRCRESAHA.116.308022.

Price, L. H., and Heninger, G. R. 1994. "Lithium in the Treatment of Mood Disorders." *The New England Journal of Medicine* 331(9):591–598. doi: 10.1056/NEJM199409013310907.

Rawat, S., Gupta, S., and Mohanty, S. 2019. "Mesenchymal Stem Cells Modulate the Immune System in Developing Therapeutic Interventions, Immune Response Activation and Immunomodulation." Rajeev K. Tyagi and Prakash S. Bisen, *IntechOpen*. doi: 10.5772/intechopen.80772.

Raza, S. S., Seth, P., and Khan, M. A. 2020. "'Primed' Mesenchymal Stem Cells: a Potential Novel Therapeutic for COVID19 Patients." *Stem Cell Reviews and Reports* 1–10. Advance online publication. doi: 10.1007/s12015-020-09999-0.

Raza, S. S., Wagner, A. P., Hussain, Y. S., and Khan, M. A. 2018. "Mechanisms Underlying Dental-derived Stem Cell-mediated Neurorestoration in Neurodegenerative Disorders. *Stem Cell Research and Therapy* 9:245. https://doi.org/10.1186/s13287-018-1005-z.

Redondo-Castro, E., Cunningham, C., Miller, J., Martuscelli, L., Aoulad-Ali, S., Rothwell, N. J., Kielty, C. M., Allan, S. M., and Pinteaux, E. 2017. "Interleukin-1 Primes Human Mesenchymal Stem Cells

Towards an Anti-inflammatory and Pro-trophic Phenotype in vitro." *Stem Cell Research Therapy* 8:79. doi: 10.1186/s13287-017-0531-4.

Ren, G., Zhang, L., Zhao, X., Xu, G., Zhang, Y., Roberts, A. I., Zhao, R. C., Shi, Y. 2008. "Mesenchymal Stem Cell-Mediated Immunosuppression Occurs via Concerted Action of Chemokines and Nitric Oxide." *Cell Stem Cell* 2(2):141–150. doi: 10.1016/j.stem.2007.11.014.

Rey, S., Luo, W., Shimoda, L. A., and Semenza, G. L. 2011. "Metabolic Reprogramming by HIF-1 Promotes the Survival of Bone Marrow-Derived Angiogenic Cells in Ischemic Tissue." *Blood* 117(18):4988–4998. doi: 10.1182/blood-2010-11-321190.

Reyes, S., Tajiri, N., and Borlongan, C. V. 2015. "Developments in Intracerebral Stem Cell Grafts." *Expert Review of Neurotherapeutics* 15(4):381–393. doi: 10.1586/14737175.2015.1021787.

Rivera-Gonzalez, G. C, Shook, B.A., Andrae, J., Holtrup, B., Bollag, K., Betsholtz, C., Rodeheffer, M. S., and Horsley, V. 2016. "Skin Adipocyte Stem Cell Self-Renewal Is Regulated by a PDGFA/AKT-Signaling Axis." *Cell Stem Cell* 19(6):738-751. doi: 10.1016/j.stem.2016.09.002.

Romanov, Y. A., Svintsitskaya, V. A., and Smirnov, V. N. 2003. "Searching for Alternative Sources of Postnatal Human Mesenchymal Stem Cells: Candidate MSC-like Cells from Umbilical Cord." *Stem Cells* 21:105–110. doi: 10.1634/stemcells.21-1-105.

Rosenzweig, H. L., Minami, M., Lessov, N. S., Coste, S. C., Stevens, S. L., Henshall, D. C., Meller, R., Simon, R. P., and Stenzel-Poore, M. P. 2007. "Endotoxin Preconditioning Protects against the Cytotoxic Effects of TNFalpha after Stroke: A Novel Role for TNFalpha in LPS-ischemic Tolerance." *Journal of Cerebral Blood Flow and Metabolism* 27(10):1663–1674. doi: 10.1038/sj.jcbfm.9600464.

Saparov, A., Ogay, V., Nurgozhin, T., Jumabay, M., Chen, W. C. W. 2016. "Preconditioning of Human Mesenchymal Stem Cells to Enhance Their Regulation of the Immune Response." *Stem Cells International* Article ID: 3924858. doi: 10.1155/2016/3924858.

Sivanathan, K. N., Gronthos, S., Rojas-Canales, D., Thierry, B., Coates, P. T. 2014. "Interferon-Gamma Modification of Mesenchymal Stem Cells: Implications of Autologous and Allogeneic Mesenchymal Stem Cell Therapy in Allotransplantation." *Stem Cell Reviews and Reports* 10:351–375. doi: 10.1007/s12015-014-9495-2.

Squillaro, T., Peluso, G., Galderisi, U. 2016. "Clinical Trials with Mesenchymal Stem Cells an Update." *Cell Transpplantation* 25(5): 829–848. doi: 10.3727/096368915X689622.

Steinberg, G. K., Kondziolka, D., Wechsler, L. R., Lunsford, L. D., Coburn, M. L., Billigen, J. B., Kim, A. S., Johnson, J. N., Bates, D., King, B., Case, C., McGrogan, M., Yankee, E. W., and Schwartz, N. E. 2016. "Clinical Outcomes of Transplanted Modified Bone Marrow-Derived Mesenchymal Stem Cells in Stroke: A Phase 1/2a Study." *Stroke* 47(7):1817–1824. doi: 10.1161/STROKEAHA.116.012995.

Sui, B. D., Hu, C. H., Zheng, C. X., Shuai, Y., He, X. N., Gao, P. P., Zhao, P., Li, M., Zhang, X. Y., He, T., Xuan, K., and Jin, Y. 2017. "Recipient Glycemic Micro-environments Govern Therapeutic Effects of Mesenchymal Stem Cell Infusion on Osteopenia." *Theranostics* 7(5):1225–1244. doi: 10.7150/thno.18181.

Sun, Y., Li, Q. F., Yan, J., Hu, R., and Jiang, H. 2015. "Isoflurane Preconditioning Promotes the Survival and Migration of Bone Marrow Stromal Cells." *Cellular Physiology and Biochemistry: International Journal of Experimental Cellular Physiology, Biochemistry, and Pharmacology* 36(4):1331–1345. doi: 10.1159/000430300.

Sun, Y., Yu, P., Zhang, G., Wang, L., Zhong, H., Zhai, Z., Wang, L., and Wang, Y. 2012. "Therapeutic Effects of Tetramethylpyrazine Nitrone in Rat Ischemic Stroke Models." *Journal of Neuroscience Research* 90(8):1662–1669. doi: 10.1002/jnr.23034.

Surugiu, R., Olaru, A., Hermann, D. M., Glavan, D., Catalin, B., and Popa-Wagner, A. 2019. "Recent Advances in Mono- and Combined Stem Cell Therapies of Stroke in Animal Models and Humans." *International Journal of Molecular Sciences* 20(23):6029. doi: 10.3390/ijms20236029.

Tang, Y., Cai, B., Yuan, F., He, X., Lin, X., Wang, J., Wang, Y., and Yang, G. Y. 2014. "Melatonin Pretreatment Improves the Survival and Function of Transplanted Mesenchymal Stem Cells after Focal Cerebral Ischemia." *Cell Transplantation* 23(10):1279–1291. doi: 10.3727/096368913x667510.

Tasaki, K., Ruetzler, C. A., Ohtsuki, T., Martin, D., Nawashiro, H., and Hallenbeck, J. M. 1997. "Lipopolysaccharide Pre-Treatment Induces Resistance against Subsequent Focal Cerebral Ischemic Damage in Spontaneously Hypertensive Rats." *Brain Research* 748(1-2):267–270. doi: 10.1016/s0006-8993(96)01383-2.

Theus, M. H., Wei, L., Cui, L., Francis, K., Hu, X., Keogh, C., and Yu, S. P. (2008)."In Vitro Hypoxic Preconditioning of Embryonic Stem Cells as a Strategy of Promoting Cell Survival and Functional Benefits after Transplantation into the Ischemic Rat Brain." *Experimental Neurology* 210(2):656–670.doi: 10.1016/j.expneurol.2007.12.020.

Toyama, K., Honmou, O., Harada, K., Suzuki, J., Houkin, K., Hamada, H., and Kocsis, J. D. 2009. "Therapeutic Benefits of Angiogenetic Gene-Modified Human Mesenchymal Stem Cells After Cerebral Ischemia." *Experimental Neurology* 216(1):47–55. doi: 10.1016/j.expneurol.2008.11.010.

Toyoshima, A., Yasuhara, T., and Date, I. 2017. "Mesenchymal Stem Cell Therapy for Ischemic Stroke." *Acta Medica Okayama* 71(4):263–268. doi: 10.18926/AMO/55302.

Tsai, L. K., Leng, Y., Wang, Z., Leeds, P., and Chuang, D. M. 2010. "The Mood Stabilizers Valproic Acid and Lithium Enhance Mesenchymal Stem Cell Migration via Distinct Mechanisms." *Neuropsychopharmacology* 35(11):2225–2237. doi: 10.1038/npp.2010.97.

Ullah, I., Subbarao, R. B., and Rho, G. J. 2015. "Human Mesenchymal Stem Cells - Current Trends and Future Prospective." *Bioscience Reports* 35(2):e00191. doi: 10.1042/BSR20150025.

Vu, Q., Xie, K., Eckert, M., Zhao, W., and Cramer, S. C. 2014. "Meta-Analysis of Preclinical Studies of Mesenchymal Stromal Cells for

Ischemic Stroke." *Neurology* 82(14):1277–1286. doi: 10.1212/WNL. 0000000000000278.

Wang, H. S., Hung, S. C., Peng, S. T., Huang, C. C., Wei, H. M., Guo, Y. J., Fu, Y. S., Lai, M. C., and Chen, C. C. 2004. "Mesenchymal Stem Cells in the Wharton's Jelly of the Human Umbilical Cord." *Stem Cells (Dayton, Ohio)* 22(7):1330–1337. doi: 10.1634/stemcells.2004-0013.

Wang, Y., Deng, Y., and Zhou, G. Q. 2008. "SDF-1Alpha/CXCR4-Mediated Migration of Systemically Transplanted Bone Marrow Stromal Cells towards Ischemic Brain Lesion in a Rat Model." *Brain Research* 1195:104–112. doi: 10.1016/j.brainres.2007.11.068.

Wei, L., Fraser, J. L., Lu, Z. Y., Hu, X., and Yu, S. P. 2012. "Transplantation of Hypoxia Preconditioned Bone Marrow Mesenchymal Stem Cells Enhances Angiogenesis and Neurogenesis after Cerebral Ischemia in Rats." *Neurobiology of Disease* 46(3): 635–645. doi: 10.1016/j.nbd.2012.03.002.

Wei, N., Yu, S. P., Gu, X., Taylor, T. M., Song, D., Liu, X. F., and Wei, L. 2013. "Delayed Intranasal Delivery of Hypoxic-Preconditioned Bone Marrow Mesenchymal Stem Cells Enhanced Cell Homing and Therapeutic Benefits after Ischemic Stroke in Mice." *Cell Transplantation* 22(6):977–991. doi: 10.3727/096368912X657251.

Wei, Z. Z., Zhu, Y. B., Zhang, J. Y., McCrary, M. R., Wang, S., Zhang, Y. B., Yu, S. P., and Wei, L. 2017. "Priming of the Cells: Hypoxic Preconditioning for Stem Cell Therapy." *Chinese Medical Journal* 130(19):2361–2374. doi: 10.4103/0366-6999.215324.

Wolf, D., and Ley, K. 2015. "Waking Up the Stem Cell Niche How Hematopoietic Stem Cells Generate Inflammatory Monocytes after Stroke." *Circulation Research* 116(3): 389–392. doi: 10.1161/CIRCRESAHA.114.305678.

Xin, H., Li, Y., Cui, Y., Yang, J. J., Zhang, Z. G., and Chopp, M. 2013. "Systemic Administration of Exosomes Released from Mesenchymal Stromal Cells Promote Functional Recovery and Neurovascular Plasticity After Stroke in Rats." *Journal of Cerebral Blood Flow and Metabolism* 33(11):1711–1715. doi: 10.1038/jcbfm.2013.152.

Yan, T., Chopp, M., and Chen, J. 2015. "Experimental Animal Models and Inflammatory Cellular Changes in Cerebral Ischemic and Hemorrhagic Stroke." *Neuroscience Bulletin* 31(6):717–734. doi: 10.1007/s12264-015-1567-z.

Yang, B., Xi, X., Aronowski, J., and Savitz, S. I. 2012. "Ischemic Stroke may Activate Bone Marrow Mononuclear Cells to Enhance Recovery after Stroke." *Stem Cells and Development* 21(18):3332–3340. doi: 10.1089/scd.2012.0037.

Yu, J., Liu, X. L., Cheng, Q. G., Lu, S. S., Xu, X. Q., Zu, Q. Q., and Liu, S. 2016. "G-CSF and Hypoxic Conditioning Improve the Proliferation, Neural Differentiation and Migration of Canine Bone Marrow Mesenchymal Stem Cells." *Experimental and Therapeutic Medicine* 12(3):1822–1828. doi: 10.3892/etm.2016.3535.

Yu, X., Chen, D., Zhang, Y., Wu, X., Huang, Z., Zhou, H., Zhang, Y., and Zhang, Z. 2012. "Overexpression of CXCR4 in Mesenchymal Stem Cells Promotes Migration Neuroprotection and Angiogenesis in a Rat Model of Stroke." *Journal of the Neurological Sciences* 316(1-2):141–149. doi: 10.1016/j.jns.2012.01.001.

Zarriello, S., Neal, E. G., Kaneko, Y., and Borlongan, C. V. 2019. "T-Regulatory Cells Confer Increased Myelination and Stem Cell Activity after Stroke-Induced White Matter Injury." *Journal of Clinical Medicine* 8(4):537. doi: 10.3390/jcm8040537.

Zhang, M., Zhao, H., Li, Z., Yang, Y., Wen, Y., Dong, J., Zhang, Q., and Ge, B. 2012. "*Effect of Salidroside on Rat Bone Marrow Mesenchymal Stem Cells Differentiation into Cholinergic Nerve Cells.*" *Chinese journal of reparative and reconstructive surgery* 26(2):158–165. https://pubmed.ncbi.nlm.nih.gov/22403877/.

Zhang, T., Gu, J., Wu, L., Li, N., Sun, Y., Yu, P., Wang, Y., Zhang, G., and Zhang, Z. 2017. "Neuroprotective and Axonal Outgrowth-Promoting Effects of Tetramethylpyrazine Nitrone in Chronic Cerebral Hypoperfusion Rats and Primary Hippocampal Neurons Exposed to Hypoxia." *Neuropharmacology* 118:137–147. doi: 10.1016/j.neuropharm.2017.03.022.

Zhang, X., Lai, W., Ying, X., Xu, L., Chu, K., Brown, J., Chen, L., and Hong, G. 2019. "Salidroside Reduces Inflammation and Brain Injury After Permanent Middle Cerebral Artery Occlusion in Rats by Regulating PI3K/PKB/Nrf2/NFκB Signaling Rather than Complement C3 Activity." *Inflammation* 42(5):1830–1842. doi: 10.1007/s10753-019-01045-7.

Zhang, Z., Zhang, G., Sun, Y., Szeto, S. S., Law, H. C., Quan, Q., Li, G., Yu, P., Sho, E., Siu, M. K., Lee, S. M., Chu, I. K., & Wang, Y. (2016). Tetramethylpyrazine nitrone, a multifunctional neuroprotective agent for ischemic stroke therapy. *Scientific reports*, 6, 37148. https://doi.org/10.1038/srep37148.

Zhao, H. B., Ma, H., Ha, X. Q., Zheng, P., Li, X. Y., Zhang, M., Dong, J. Z., and Yang, Y. S. 2014a. "Salidroside Induces Rat Mesenchymal Stem Cells to Differentiate into Dopaminergic Neurons." *Cell Biology International* 38(4):462–471. doi: 10.1002/cbin.10217.

Zhao, H. B., Qi, S. N., Dong, J. Z., Ha, X. Q., Li, X. Y., Zhang, Q. W., Yang, Y. S., Bai, J., and Zhao, L. 2014b. "Induces Neuronal Differentiation of Mouse Mesenchymal Stem Cells Through Notch and BMP Signaling Pathways." *Food and Chemical Toxicology: an International Journal published for the British Industrial Biological Research Association* 71:60–67. doi: 10.1016/j.fct.2014.05.031.

Zhao, L., Hu, C., Han, F., Cai, F., Wang, J., and Chen, J. 2020. "Preconditioning is an Effective Strategy for Improving the Efficiency of Mesenchymal Stem Cells in Kidney Transplantation." *Stem Cell Research and Therapy* 11(1):197. doi: 10.1186/s13287-020-01721-8.

Zhao, M. Z., Nonoguchi, N., Ikeda, N., Watanabe, T., Furutama, D., Miyazawa, D., Funakoshi, H., Kajimoto, Y., Nakamura, T., Dezawa, M., Shibata, M. A., Otsuki, Y., Coffin, R. S., Liu, W. D., Kuroiwa, T., and Miyatake, S. 2006. "Novel Therapeutic Strategy for Stroke in Rats by Bone Marrow Stromal Cells and ex vivo HGF gene Transfer with HSV-1 Vector." *Journal of Cerebral Blood Flow and Metabolism* 26(9):1176–1188. doi: 10.1038/sj.jcbfm.9600273.

Zhong, Z., Han, J., Zhang, J., Xiao, Q., Hu, J., and Chen, L. 2018. "Pharmacological Activities, Mechanisms of Action, and Safety of

Salidroside in the Central Nervous System." *Drug Design, Development and Therapy* 12:1479–1489. doi: 10.2147/DDDT.S160776.

Zhou, L., Lin, Q., Wang, P., Yao, L., Leong, K., Tan, Z., and Huang, Z. 2017. "Enhanced Neuroprotective Efficacy of Bone Marrow Mesenchymal Stem Cells Co-overexpressing BDNF and VEGF in a Rat Model of Cardiac Arrest-induced Global Cerebral Ischemia." *Cell Death and Disease* 8(5):e2774. doi: 10.1038/cddis.2017.184.

Zhou, L., Yao, P., Jiang, L., Wang, Z., Ma, X., Wen, G., Yang, J., Zhou, B., and Yu, Q. 2020. "Salidroside-Pretreated Mesenchymal Stem Cells Contribute to Neuroprotection in Cerebral Ischemic Injury in vitro and in vivo." *Research Square* 2020, Version 1. doi: 10.21203/rs.3.rs-59580/v1.

Zin'kova, N. N., Gilerovich, E. G., Sokolova, I. B., Shvedova, E. V., Bilibina, A. A., Krugliakov, P. V., and Polyntsev, D. G. 2007. "Mesenchymal Stem Cells Transplantation Influences upon Dynamics of Morphological Changes in Rat Brain after Stroke." *Tsitologiia* 49(11):923–932. https://pubmed.ncbi.nlm.nih.gov/18217359/.

Zuk, P. A., Zhu, M., Ashjian, P., De Ugarte, D. A., Huang, J. I., Mizuno, H., Alfonso, Z. C., Fraser, J. K., Benhaim, P., and Hedrick, M. H. 2002. "Human Adipose Tissue is a Source of Multipotent Stem Cells." *Molecular Biology of the Cell* 13(12):4279–4295. doi: 10.1091/mbc.e02-02-0105.

Zuo, W., Yan, F., Zhang, B., Hu, X., and Mei, D. 2018. "Salidroside Improves Brain Ischemic Injury by Activating PI3K/Akt Pathway and Reduces Complications Induced by Delayed tPA Treatment." *European Journal of Pharmacology* 830:128–138. doi: 10.1016/j.ejphar.2018.04.001.

Chapter 3

STARVATION RATIONS: THE THERAPEUTIC POTENTIAL OF KETONE BODIES FOR STEM CELL FUNCTION

Mary Board, DPhil
St. Hilda's College, University of Oxford, Oxford, UK

ABSTRACT

Ketone bodies, produced primarily by the liver and, to a lesser extent, by intestinal epithelial cells, have traditionally been regarded as a fuel of starvation which allows the body to integrate stores of fat with those of carbohydrate and to deploy glucose-sparing mechanisms. More recently, renewed scrutiny of these starvation substrates has been stimulated by observations of the neuroprotective and anti-cancer effects of the ketogenic diet and the anti-aging consequences of caloric restriction. Ketone bodies, D-3-hydroxybutyrate (HB) and acetoacetate, have multifarious effects in addition to their roles as oxidative, energy-yielding substrates. Signalling through G-protein coupled receptors serves to integrate fat and carbohydrate metabolism, promoting starvation-appropriate substrate-selection to achieve a glucose-sparing effect. Consumption of a ketone body substrate also reduces production of reactive oxygen species (ROS) in many cell-types. Reduced rates of ROS-production may drive the preference shown by human mesenchymal stem cells for a ketone body substrate, with consumption rates being up to 35-fold higher than those of glucose, accompanied by a 45-fold suppression of ROS-production. Selection of HB as an oxidative substrate, along with other properties of stem cells, such as the residence in hypoxic niches within the body, may reduce the oxidative damage, including mutation and accelerated apoptosis, that accompany oxidative stress. HB has multifarious effects on gene expression including inhibitory activity towards Class I and Class IIa histone deacetylases and an ability to cause hydroxybutyrylation of histone proteins. As a result, expression of genes which reduce oxidative stress is promoted. Expression and activity of cell cycle regulators are also affected, contributing to observed suppression of proliferation and maintenance of stemness in intestinal stem cells and influencing the processes of differentiation and apoptosis. Such observations regarding the impact of ketone bodies on stem cells have implications for stem cell therapies which have been promoted for neurodegenerative disorders and for myocardial damage, among other conditions. This review assesses the potential benefits of *in vitro* culture of therapeutic stem cells in the presence of HB along with the ketogenic diet, whereby higher physiological concentrations of ketone bodies can be achieved *in vivo*, as an adjuvant to stem cell transplantation. Thus, the ketogenic diet, long recognised for its ability to ameliorate refractory epilepsy and its favourable impact on neurological function and more recently proposed as an anti-cancer therapy, may have renewed applications as an accompaniment to stem cell therapy.

Keywords: mesenchymal stem cells, ketone bodies, D-3-hydroxybutyrate, ROS, metabolism

KETONE BODIES: SIGNIFICANT SUBSTRATES FOR STEM CELLS

The therapeutic potential of stem cells of different origins has caused much excitement in recent years. The list of conditions considered appropriate targets for stem cell therapy has grown with the initial successes surrounding trial treatments of Parkinson's Disease and of cardiac tissue remodelled by myocardial infarction giving much encouragement. A variety of stem cells have been suggested as candidates for therapy, including mesenchymal stem cells, and the potential for induced pluripotential stem cells (IPSCs) as a means of reducing immune stimulation is also under investigation. Stem cells destined for therapeutic implantation may have their origins in the recipient's tissues, a donor's tissues or may be induced *in vitro*. Cells from all sources are likely to be maintained *in vitro* for an episode of storage and/or culture. During their *in vitro* maintenance, the conditions are crucial to ensure that the cells retain their stem-like characteristics, avoiding differentiation and reducing progression of the stem cell population to apoptosis both of which processes would reduce the number of viable cells available for tissue regeneration in the recipient. Once transplanted, the cells must also retain their stem-like characteristics until *in situ* in the target tissues. This review attempts to assess the evidence for beneficial adjustments to the culture medium of mesenchymal and other stem cells and to the *in vivo* conditions surrounding implantation of therapeutic stem cells that would optimise conditions for repair and regeneration. Our work with human mesenchymal stem cells (hMSCs) has indicated that cultured cells undergo metabolic changes at low passage number before alterations in markers of stemness can be detected. Inclusion of ketone bodies (KBs) in the culture medium prevents some of the culture-dependent changes (Board, 2017) and may have desirable effects for the stem cell population.

KBs, acetoacetate (AA) and D-3hydroxybutyrate (HB) represent efficient and accessible oxidative substrates which serve to integrate fat and carbohydrate metabolism on a whole body basis, achieving a glucose-sparing effect. These energy-yielding substrates have abundant signalling roles to indicate the fasted/exercising/energy-deprived state and coordinate the responses of the body's tissues appropriately. These properties make KBs of potential significance to dividing cells, including stem cells. The levels of KBs in the plasma would be expected to indicate whether or not the body's energy status is sufficient to support cell division and oxidation of these fuels generates fewer damaging reactive oxygen species (ROS) which might otherwise contribute to the accumulation of mutations during cell division. Moreover, significant signalling roles have now been ascribed to ROS, themselves, which indicate a potential impact on rates of cell proliferation. Recent findings on fuel utilisation by cultured human mesenchymal stem cells (hMSCs) indicate that these cells consume KBs in preference to other fuels, such as glucose, and that the KBs are fully oxidised to CO_2. The molar rates of oxidation of AA outstrip those of glucose by a factor of up to 35 and are sufficient to synthesise at least 17 times as much ATP (Board, 2017). As well as implying that fuel utilisation by hMSCs can be substantially oxidative, rather than the anaerobic, glycolytic metabolism that has sometimes been ascribed to these cells (Patappa, 2011; Simsek, 2010), these findings imply that KBs may be a significant contributor to ATP-generation by hMSCs, a cell-type expected to have a high ATP requirement in order to fuel the processes of proliferation and differentiation.

There are reasons to suspect that the influence of KBs on stem cells might extend to the *in vivo* situation. There are many significant and convincing reports of neuroprotective effects of KBs, ranging from the long-established ketogenic diet (KGD) for the treatment of epilepsy, to the more recent investigations of the potential for amelioration of neurodegenerative disorders, such as Parkinson's Disease, Alzheimer's Disease and the traumatic condition of stroke. It is no coincidence that the latter group of disorders are also targets for potential stem cell therapy and, in the case of Parkinson's Disease, the results of early trials seem to be especially promising. These observations expose the possibility that a combination of stem cell transplantation and administration of KBs might be a successful therapeutic approach. The potential links between stem cells and KBs illustrate that a vital next step is to assess the impact of these substrates on stem cells on both a molecular and on a whole body level.

ROLES OF KETONE BODIES

KBs comprise two four carbon breakdown products of fatty acid metabolism, AA and HB. The three carbon compound, acetone, is sometimes included in the group but represents a minor component of the plasma although it can achieve concentrations of 0.76 mM when AA is at high concentrations after a 3-day fast (Riechard, 1979) and loses a carbon atom by a spontaneous process of decarboxylation in order to contribute to acid-base balance and mitigate the pH-lowering effects of high AA plus HB concentrations. Roles for acetone are uncertain and its volatility means that it is substantially excreted through the breath, although it may also be metabolised to energy-yielding substrates. HB and AA are transported in the bloodstream in a ratio that varies between 1HB:1AA in the normal fed state to 4HB:1AA in the normal fasted state (Cahill, 1966) and can achieve values of 5HB:1AA in diabetic ketoacidosis (Stephens, 1971). Both compounds are weak acids with PKas of 3.6 mM (AA) and 4.7 mM (HB) so that high concentrations can affect the pH of the blood compartment. KBs are transported in the blood in their free state, unlike fatty acids which must be complexed with proteins to increase their solubility. Hence, whereas tissues such as neurones are unable to take up fatty acids due to the absence of lipoprotein receptors, these tissues may consume KBs since they express the monocarboxylate group of transporters. AA and HB are interconverted by the action of D-3-HB dehydrogenase in target cells in which AA is the substrate for the ketolytic pathway and energy-yielding form.

ENERGY-YIELDING SUBSTRATES

The ketogenic pathway is expressed by the liver, which tissue has a high capacity for KB output capable of producing up to 300g per day (Ballasse, 1989) and is the primary source of these substrates, and intestinal epithelial cells. The pathway is probably not present to any great extent in other tissues (reviewed by Puchalska, 2017). The liver's high capacity for ketone body output allows this tissue to integrate fat and carbohydrate metabolism during starvation and supply the central nervous system (CNS) with an oxidisable substrate. The brain uses only glucose in the fed state and supplies its large energy requirements (20% of resting energy expenditure in the fat free mass of the adult) by oxidising this substrate, accounting for a glucose consumption rate of approximately 120g per day. Whereas the CNS cannot take up fatty acids as an oxidative substrate, during starvation the brain gradually adapts to replacing some of its glucose requirement with KBs and can derive up to 75% of its energy needs from this substrate from about 3 days of starvation (in the adult). The brain does retain a requirement for glucose oxidation and at least 25% of its energy must come from this source even during prolonged starvation. Oxidation of KBs by the CNS under starvation conditions requires a supply of the tricarboxylic acid cycle intermediate, oxaloacetate, with which acetyl CoA must condense in order to achieve full oxidation (Figure 1). Oxaloacetate must derive from a carbohydrate source since no pathway between acetyl CoA and oxaloacetate exists in mammalian cells. However, the use of KBs as an energy source by the brain, among other tissues, reduces the dependence of this tissue on blood glucose and allows fat deposits to contribute to the brain's energy requirement. The replacement of some of the brain's glucose requirement by KBs reduces the burden on hepatic gluconeogenesis, the carbon source for which must draw on muscle protein content. This adaptation has the effect of protecting muscle mass and delaying the loss of function of essential cardiac and diaphragm muscles due to wasting, usually the eventual cause of starvation-related death. Thus, in whole body terms, KBs perform a vital glucose-sparing role in integrating fat and carbohydrate metabolism. KBs are an efficient ATP-generating substrate: yields are 22.5 moles ATP per mole HB (5.6 moles ATP per carbon atom) compared with 31 moles ATP per mole glucose (5.2 moles ATP per carbon atom). With respect to hMSCs, AA is a preferred ATP-generating substrate, oxidised at up to 35 times the rate of glucose. We have found that supplying cultured hMSCs with exogenous pyruvate in addition to AA increases the rates of AA oxidation 1.5-fold and thus also increases calculated ATP-yields (Board, 2017). Cells *in vivo* must utilise a carbohydrate source, such as pyruvate, to generate oxaloacetate and maximise the efficiency of ketone body oxidation (Figure 1). Given the high rates of expression of UCP2 in these cells, which may shunt glycolytic pyruvate out of the mitochondrion, the source of pyruvate is unlikely to be glucose (Bouillaud, 2009; Zhang, 2011). However, there is evidence that stem cells *in vivo* may take up pyruvate released by other cells into the stroma (Alt, 2005; O'Donnell-Tormey, 1987), indicating a degree of inter-cell cooperation in substrate selection. In addition to their roles as energy-yielding substrates, there is established and emerging evidence that oxidation of KBs by mesenchymal stem cells and other cell-types generates fewer damaging reactive oxygen species (ROS) than the equivalent amount of glucose (Board, 2017; Quinlan, 2013).

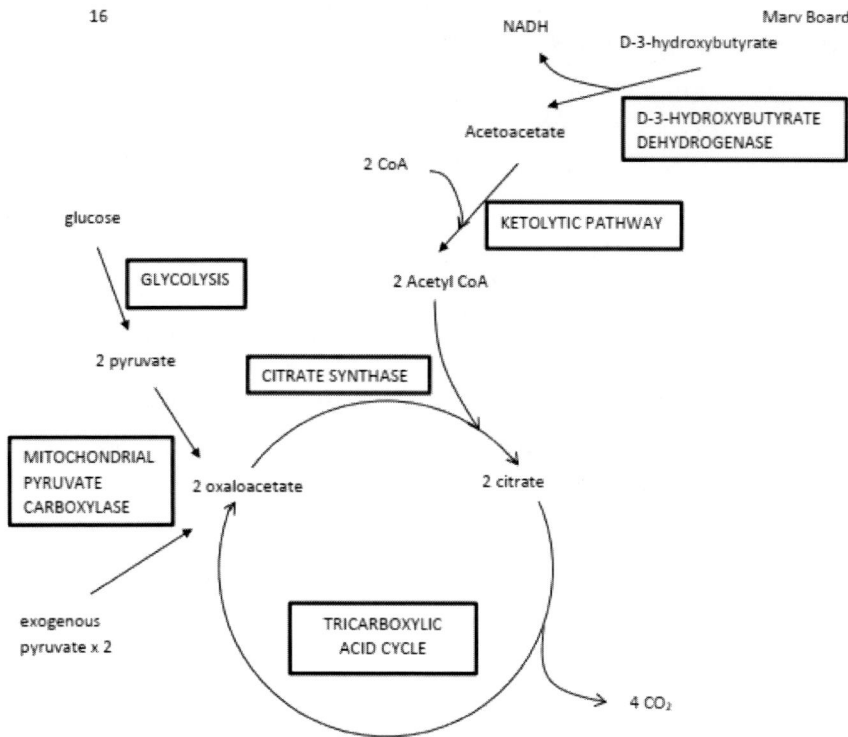

Figure 1. Pathway of metabolism of D-3-hydroxybutyrate and acetoacetate. Ketone bodies are metabolised to acetyl CoA by the ketolytic pathway and then may condense with oxaloacetate to form citrate and be converted to CO_2 by reactions of the tricarboxylic acid cycle. Oxaloacetate is required to enable oxidation of acetoacetate-derived acetyl CoA by the tricarboxylic acid cycle. Oxaloacetate must derive from a carbohydrate source, such as glucose (converted to pyruvate by reactions of glycolysis) or exogenous pyruvate.

PRODUCTION OF ROS

ROS comprise a group of unstable, highly reactive products of oxygen metabolism that have an unpaired electron and cause damage to cellular macromolecules, including DNA and lipids, by forming addition products. It is considered that the most significant sites of ROS-production are complexes I and III of the electron transport chain when electrons are transferred to oxygen from electron carriers. The superoxide anion, $O_2^{\cdot}$, and the hydroxyl radical, $OH^{\cdot}$, are interconverted with hydrogen peroxide by the cellular machinery and the radicals are targeted by the antioxidant system including superoxide dismutase (SOD), catalase and glutathione peroxidase. The complexity of the antioxidant machinery indicates the effort put by the cell into reducing the impact of ROS and these have traditionally been considered destructive by-products of oxygen metabolism that contribute to the accumulation of mutations and accelerate the ageing process. The accumulation of ROS within the cell indicates a state of oxidative stress and impending damage. Thus, it is not surprising that recent research has indicated signalling roles whereby the progression of the cell cycle is promoted by high ROS levels (reviewed by Redza-Dutortoir, 2016) resulting in increased proliferation of stem cells (Hole, 2010; Chowdury, 2010) and progression to apoptosis. Proliferation may be stimulated via a signalling pathway which utilises Yap protein in combination with specific transcription factors (Ji, 2017). Prolonged exposure to high levels of ROS promotes apoptosis (Simon, 2000; Circu, 2010) and may contribute to an increased accumulation of mutations during proliferation (Park, 2020; Han, 2018; Wang, 2017(a)). In the specific case of stem cells, ROS signal via transcription factors, including FOXOs, and affect cell cycle progression (LeBelle, 2011). Reduced levels of ROS appear to cause cell cycle arrest and delay the progression of the stem cell population to differentiation. It has been observed that the occupation of hypoxic niches by stem cells within the body may be an adaptation to minimise the generation of ROS plus stem cells seem to have an enhanced anti-oxidant capacity relative to their differentiated counterparts (Madhavan, 2006). Both these observations indicate that maintenance of a low ROS concentration within the stem cell population is significant for function, perhaps assisting with the retention of stemness. Thus, exposure to high levels of ROS may have a significant role in stimulating rates of proliferation and apoptosis.

Many observations have been made to the effect that consumption of KBs by cells, reduces the production of ROS (Maalouf, 2009; Rojas-Morales, 2020). AA consumption by hMSCs reduces the production of ROS 45-fold relative to glucose consumption and may explain why these cells *in vitro* show a marked preference for an AA substrate, consuming it at rates up to 35-fold greater than those of glucose (Board, 2017). HB induces genes to curtail oxidative stress, including catalase and SOD, in human embryonic kidney cells (Shimazu, 2013) and in spinal cord (Kong, 2017). The transcription factor, Nrf2, is a master regulator of the cell's ability to sense oxidants and coordinate the response. When activated, Nrf2 controls the transcription of a suite of genes to reduce oxidative stress (reviewed by Ma, 2013) and regulates many anti-oxidant responses of the cell. The activity of Nrf2 is sufficiently potent for it to be investigated as a pharmacological target for the relief of oxidative stress and the accompanying inflammatory response (Robledinos-Anton, 2019). Butyrate is known to activate Nrf2 (Dong, 2017) and it may be that KBs also stimulate this transcription factor since activity is known to be upregulated in response to the KGD (Cavaleri, 2018).

The occupation by stem cells of hypoxic niches within their respective tissues has led to many observations of anaerobic or glycolytic metabolism in these cell populations. *In vitro*, when exposed to normoxic conditions, the same types of stem cells demonstrate a high capacity for oxidative metabolism, indicating that the required pathways are expressed and active within these cells. The adoption of a hypoxic environment may be a protective mechanism for stem cells in order to reduce the rates of production of ROS that accompany oxidative metabolism in order to reduce the possibility of oxidative stress and to maintain stem-like properties and lower rates of proliferation within the stem cell population. There is emerging evidence that the preference shown by stem cells for HB as an oxidative substrate may contribute to the anti-oxidant protection mechanisms of the cellular defence machinery. UCP2 is known to be widely expressed among various stem cell types and may play a role in substrate selection by these cells.

UCP2 AS A REGULATOR OF ROS

UCP2 is a member of the anion carrier superfamily, is expressed in the inner mitochondrial membrane of a wide range of cell-types (Ricquier, 2000) and has 59% homology to UCP1 (Fleury, 1997). UCP1 is known to mediate mitochondrial uncoupling and thermogenesis in brown adipose tissue. UCP2 is widely expressed in stem cells from various origins, including mesenchymal stem cells (Zhang, 2017; Board, 2017). Although its function is not fully characterised, it is thought to influence pyruvate transport into the mitochondrion (Bouillaud, 2009; Diano, 2012) and to mediate the exchange of intermediates of the tricarboxylic acid cycle for phosphate, thus regulating oxidation of glucose-derived carbon (Vozza, 2014). Gene dosage experiments have also demonstrated that UCP2 reduces the membrane potential of the inner mitochondrial membrane in rat muscle (Marti, 2001), perhaps because it acts as a mild uncoupler. Inhibition of UCP2 activity by the specific inhibitor, genipin, leads to increased production of ROS by hMSCs (Board, 2017), indicating a role for this protein in suppressing ROS-production. Whether UCP2 reduces ROS-production directly due to its uncoupling action or whether the effect is due to reduced rates of glucose oxidation when UCP2 is active, since oxidation of this substrate is known to stimulate ROS-production (Board, 2017), is not clear. Whatever the mechanism involved, a role has been suggested for UCP2 allowing it to regulate the activity of ROS in stem cells, an effect which might contribute to the retention of stem-like characteristics by the cells. Indeed, it has been suggested that increased activity of UCP2 may influence differentiation and reduce rates of apoptosis in stem cells (Zhang, 2011). Apoptosis is a process responsible for significant losses of transplanted stem cells exposing UCP2 as a potential target to increase rates of stem cell delivery during therapy. It has been suggested that UCP2 may serve to direct metabolism within the stem cell mitochondrion away from glucose oxidation and towards glutaminolysis (Vozza, 2014). We consider that UCP2 may also direct stem cell mitochondrial metabolism towards ketolysis if ketone body substrates are available. Replacing the glucose substrate for these cells both reduces ROS-production and protects the stem-like status of the cells. Given that HB is known to stimulate production of UCP2 protein both in cultured nerve cells and in hippocampal neurones of mice fed a KGD (Hasan-Olive, 2019), the same effect of HB on expression of UCP2 in mesenchymal and other stem cells is possible. Given the potential for therapeutic use of these cells, investigations into the stimulation of UCP2 activity via HB levels commensurate with the KGD or via another mechanism would provide a potentially productive avenue of enquiry.

SIGNALLING BY KETONE BODIES

Traditionally, KBs have been considered fuels of intermediate and prolonged starvation since plasma concentrations in the fed state are low at about 0.1 mM (Robinson,1980) rising to 1-2 mM after 1-2 days of starvation (Cahill, 1966; Robinson, 1980) and 6-8 mM after prolonged starvation (Cahill, 2006). Concentrations of 1-2 mM are achieved after 90 minutes of exercise (Koeslag, 1980), a situation, like starvation, which puts an added burden on energy-yielding substrates. Given this observation, it is not surprising that a number of signalling roles have been identified for HB, in particular. HB is known to antagonise GPCR41 (Kimura, 2011) and suppress sympathetic nervous system activity to decrease whole body energy expenditure. HB also binds GPCR109A, expressed in adipocytes and immune cells. By activating GPCR109A in the adipocyte, HB suppresses the activity of hormone-sensitive lipase (Taggart, 2005) in order to reduce rates of lipolysis and supply of substrates for ketogenesis, a negative feedback mechanism. Activation of the same receptor in immune cells is considered to have an anti-inflammatory effect (Wu, 2010; Digby, 2012). It is noteworthy that effects of HB on GPCR109A evince an IC_{50} of 0.7 mM (Taggart, 2005), a concentration that would be achieved after a short period of fasting. More recently, it has been speculated that signalling by AA at GPCR43 may mediate fasting and have an impact on neuronal function (Miyamoto, 2019). Overall, the signalling roles of AA and HB appear to coordinate responses of the body to ketogenic conditions. HB also has effects on gene expression due to its inhibition of both Class I and Class IIa histone deacetylases (HDACs) (Shimazu, 2013), increasing the degree of acetylation of histone proteins and fulfilling an epigenetic function. Moreover, hydroxybutyrylation of histones on lysine residues shows a potent increase in the presence of HB, an effect which may promote the expression of starvation-dependent genes (Xie, 2016). Both receptor-dependent signalling and epigenetic effects of KBs are essential indicators of the starved or carbohydrate-deprived state and serve to coordinate the body's response to such states. The abundance of fuels within the body might be expected to have an impact on dividing cells, indicating whether sufficient resources are available for proliferation. These effects concern the limited variety of cells within the body which retain a capacity for division, including lymphocytes and epithelial cells and the stem cells resident within tissues.

EFFECTS OF KETONE BODIES ON CELL DIVISION

There have been many reports of KBs suppressing cell proliferation in a diverse array of cell-types. The culture of bovine lymphocytes in the presence of HB suppresses proliferation (Sato, 1995) and levels of acetate commensurate with ketosis also inhibit bovine lymphocyte proliferation (Franklin, 1991). Moreover, KBs alone do not support proliferation of cultured rat lymphocytes (Ardawi, 1984). Such observations may represent the limitations of KBs in the absence of an appropriate carbohydrate source as energy-yielding substrates for these cells in culture or there may be more specific signalling and gene transcription effects involved. Effects are also strongly concentration-dependent. Mice fed a KGD had improved responses to infection with flu virus, indicating that exposure to physiological concentrations of KBs in themselves does not depress the immune response (Pardo, 2020; Goldberg, 2019). Interestingly, the improvement in immune response shown by mice fed a KGD, which included increased numbers of γδT cells in lung, was not reproduced in mice fed HB alone (Goldberg, 2019). This might imply that a KGD, which stimulates β-oxidation of fats and ketogenesis, may stimulate the immune response but high levels of KBs, commensurate with the state of more pronounced or prolonged ketosis, may depress the same response. Such an observation might explain why diabetics and endurance athletes, both groups considered to be exposed to consistently high levels of plasma ketones, tend to suffer higher rates of infections than the general population.

The relationship between intestinal epithelial cells and KBs has received a great deal of research attention as a result of observations that intestinal cells can use butyrate produced from gut bacteria to synthesise KBs and also that the intestinal epithelium has a capacity for regeneration due to its content of stem cells. Failure of the intestinal epithelium to regenerate appropriately, due to abnormal cycles of proliferation, differentiation and apoptosis of intestinal epithelial stem cells, is associated with a suite of conditions including irritable bowel syndrome, necrotising enteroiditis and colorectal cancer. There have been many reports that both butyrate and HB promote cell cycle arrest and inhibit proliferation of intestinal stem cells. The cells require Notch activity in order to maintain stemness and activity is promoted by a dynamic cycle of HB-production and action (Cheng, 2019). Indeed, knockout of the gene encoding HMGCoA synthetase 2, considered rate-limiting for HB-production, reduced the population of intestinal stem cells and administration of HB restored regenerative capacity (Cheng, 2019). HB may stimulate Notch via its inhibition of HDAC activity. Thus, overall intestinal integrity may rely on the cycle of production of ketone bodies from butyrate followed by the signalling actions of HB. Thus, the potential for the KGD or external administration of KBs to treat intestinal disorders associated with abnormal regeneration of the intestinal epithelium are being scrutinised. Such reports regarding functional stem cells of the body where HB may have a role in regulating rates of proliferation have stimulated interest in the effects of the same substrate on proliferation of tumour cells. There have been many reports of HB suppressing the growth of malignant cells in, for example, breast cancer (Bartmann, 2018), neuroblastoma (Skinner, 2009), colon cancer (Fine, 2009) and pancreatic cancer (Shukla, 2014). These observations have led to the suggestion that genes encoding enzymes of ketone body metabolism may be identified as "tumour suppressor genes" (Cui, 2019). Thus, the KGD has received renewed scrutiny for its potential anti-cancer effects (Weber, 2018) to add to its roles in neuroprotection and promotion of intestinal integrity.

HB, along with the structurally related butyrate, is well-established as a regulator of gene transcription via its role as an inhibitor of HDAC activity. HDAC activity is required for gene expression and reduced rates of expression are associated with the hyperacetylation promoted by HB, AA and butyrate. Both class I and IIa HDACs are inhibited by HB with an IC_{50} of around 2-5mM (Wang, 2017(b)), concentrations commensurate with ketosis due to intermediate or prolonged starvation or to the KGD. Interestingly, proteins other than histones are substrates for HDACs, including the cell cycle regulators, p53 and c-myc. Thus, the cumulative evidence seems to suggest that higher, physiological concentrations of KBs may have a generalised effect in promoting cell cycle arrest and inhibiting cell proliferation. If the same findings prove to be true in the case of all stem cells, whether endogenous or exogenous, indications are that ketone body concentrations associated with moderate starvation or with the KGD may promote the retention of stem-like properties in the stem cell population. In the situation of stem cell therapy, cells must be stored *in vitro* and often cultured to increase cell number prior to implantation in the recipient. Culture of the cells in the presence of KBs plus higher plasma levels of the same substrates in the recipient may have the effect of promoting stem-like properties and increasing the length of time the cells remain in the stem-like state. This would increase the efficiency of targeting of cells to the recipient tissue enhancing the probability that transplanted cells are *in situ* before differentiation commences. In addition, the influence of HB on the cell cycle has the consequence of reducing rates of apoptosis, a major cause of loss of stem cell number during transplantation. Thus, there are many positive indications of the potential for a combination of the KGD with stem cell therapy.

THE KETOGENIC DIET

The KGD has been in use at least since the 1920s (Wilder, 1921; Peterman, 1924) for the treatment of refractory epilepsy in children and adults (reviewed by D'Andrea-Meira, 2019 and by Hartmann, 2007). The classical KGD consists of a weight for weight ratio of fat to carbohydrate plus protein of 4:1 with the result that 90% of consumed calories derive from fat (Kossof, 2009) and is highly successful in reducing seizure frequency and duration. The diet generally aims to maintain plasma KB concentrations above 2 mM (Kim, 2008) although significantly higher blood concentrations of 5-8 mM have been reported during trials using the classical KGD (Neal, 2009; Musa-Veloso, 2006). The reasons for the success of the treatment regimen are still unclear, although it has been speculated that KBs may suppress the release of the excitatory neurotransmitter, glutamate, from synaptic vesicles (Juge, 2010), and increase effects of the inhibitory neurotransmitter, GABA (Daikhin, 1998). Moreover, both AA and HB have an effect on K^+ATP channels which may limit the transmission of seizures (Ma, 2007) and UCPs are known to be upregulated by the KGD which suppresses ROS-production (Sullivan, 2004), reducing oxidative stress in neurones. It is probably for these long-recognised neuroprotective effects that the KGD has recently attracted attention as a beneficial component of treatment for neurodegenerative disorders. The KGD has been shown to ameliorate the conditions of Parkinson's Disease (Philips, 2018), Alzheimer's Disease (Reger, 2004) and stroke (Gibson, 2012) in animal models and human patients and has stimulated discussion on the neuroprotective properties of KBs. It is no coincidence that neurodegenerative disorders are a group of conditions that are currently under investigation as the targets for stem cell therapy. Investigations are advanced with respect to Parkinson's Disease (Barker, 2017) and progress in animal models shows great promise for Alzheimer's disease (reviewed by Zhang, 2020) and for stroke (reviewed by Borlongan, 2019). Observations that stem cell therapy and KGD might be combined as a treatment for neurological disorders make it a pressing concern to assess the molecular effects of KBs on stem cells. Although a great deal of research remains to be done, initial findings imply that higher physiological concentrations of KBs may reduce rates of proliferation and promote the retention of stem-like properties in stem cell populations which would be of huge significance to recipients of stem cell transplants and may protect transplanted cells from apoptosis, increasing the time available for targeting of cells and the chances of the cells being *in situ* before differentiation takes place. There is also much evidence to support the reduction of oxidative stress by stem cells in the presence of higher physiological concentrations of KBs which may assist the cells with their native efforts to minimise ROS-production. Following the interest in the role of cancer stem cells in tumorigenesis and tumour promotion, any impact of KBs on rates of tumour proliferation is also of current interest and recent findings indicate an inhibitory effect of KBs on tumour cell proliferation.

THE POTENTIAL OF HB AND THE KGD IN STEM CELL THERAPY

The current state of knowledge regarding the impact of KBs on human stem cells indicates a considerable potential for combining a ketone body based adjuvant approach to the therapeutic transplantation of stem cells. The approach would involve both the *in vitro* culture and storage of cells in the presence of physiological HB concentrations in the pre-implantation stage and the use of the KGD to maintain higher physiological KB concentrations during post-implantation therapy. Indications from use of cultured cells and from animal studies are that higher, physiological concentrations of HB suppress stem cell proliferation and delay the progression of the cells to differentiation and apoptosis. Similar concentrations of HB also suppress ROS-production by the cells with the effect of reducing oxidative stress and potential rates of mutation. Pre-implantation culture and storage of cells in the presence of HB may have the effect of extending the scope of this phase by minimising loss of stemness by the cell population. Enrichment of stem cell culture medium with HB would seem to be an approach worthy of further investigation. Previous work by our group has demonstrated that culture of hMSCs in the presence of HB prevents some of the culture-dependent metabolic changes that otherwise take place at low passage number (Board, 2017). In the post-implantation stage of stem cell therapy, the presence of higher physiological ketone body concentrations would promote the retention of stemness by the cell population within the body. Higher physiological HB concentrations can be achieved *in vivo* by use of the KGD. Delay of stem cell differentiation would allow more opportunity for targeting of appropriate numbers of stem cells to the relevant tissue before differentiation for repair and regeneration takes place. HB seems likely to reduce the progression of the stem cell population to apoptosis and cell loss, increasing the numbers available for regeneration. Indeed, many of the long-established neuroprotective effects of the KGD may be due to similar mechanisms by minimising oxidative stress in neurones and reducing the rates of neuronal apoptosis that contribute to neurodegeneration. Effects of HB are sensitive to concentration since it is known that supraphysiological concentrations, such as may be encountered in diabetic ketosis or ketoacidosis, are teratogenic (Horton, 1983; Moore, 1989;) and otherwise harmful. However, judicious use of the KGD is not considered to put subjects at risk of hyperketonaemia. The promising work to date indicates a great deal of potential for the use of the *in vitro* culture of stem cells with HB and of the KGD as adjuvant therapies to stem cell transplantation. In the case of neurodegenerative disorders where the approach towards stem cell therapy is already advanced, the KGD is known to have a pre-established neuroprotective effect.

REFERENCES

Alt R., Riemer T., Fiehn O., Niederwieser D. and Cross M. (2005) Evidence for restricted glycolytic metabolism in primary CD133+ cells. *Blood* 106: (Abs. 1726).

Ardawi, M. S. M. and Newsholme, E. A. (1984) Metabolism of ketone bodies, oleate and glucose in lymphocytes of the rat *Biochem J* 221 (1): 255-60.

Ballasse, E. O. and Féry, F. (1989) Ketone body production and disposal: effects of fasting, diabetes, and exercise *Diab Met Rev* 5: 247-70.

Barker, R. A., Parmar, M., Studer, L. and Takahashi, J. (2017) Human Trials of Stem Cell-Derived Dopamine Neurons for Parkinson's Disease: Dawn of a New Era *Cell Stem Cell* 21: 569–573.

Board, M., Lopez, C., van den Bos, C., Callaghan, R., Clarke, K. and Carr, C (2017) Acetoacetate is a more efficient energy-yielding substrate for human mesenchymal stem cells than glucose and generates fewer reactive oxygen species *Int J Biochem Cell Biol,* 88: 75-83.

Bouillaud F. UCP2, not a physiologically relevant uncoupler but a glucose sparing switch impacting ROS production and glucose sensing (2009) *Biochim. Biophys. Acta.* 1787:377–383.

Borlongan, C. V. (2019) Concise Review: Stem Cell Therapy for Stroke Patients: Are We There Yet? *Stem Cell Transl Med* 8 (9): 983-8.

Bartmann, C., Raman, S. R. J., Floter, J., Schulze, A., Bahlke, K., Willingstorfer, J., Strunz, M., Wockel, A., Klement, R. J., Kapp, M., Djuzenova, C. S., Otto, C. and Kammerer, U. (2018) Beta-hydroxybutyrate (3-OHB) can influence the energetic phenotype of breast cancer cells, but does not impact their proliferation and the response to chemotherapy or radiation *Cancer & Metabolism* 6: Article 8.

Cahill, G. F., Herrera, M. G., Morgan, A. P., Soeldner, J. S., Steinke, J., Levy, P. L., Reichard, G. A. and Kipnis, D. M. (1966) Hormone-fuel interrelationships during fasting. *J Clin Invest,* 45: 1751-69.

Cahill, G. F. (2006) Fuel Metabolism in starvation *Ann Rev Nut* 26 (1): 1-22.

Cavaleri, F. and Bashar, E. (2018) Potential Synergies of β-Hydroxybutyrate and Butyrate on the Modulation of Metabolism, Inflammation, Cognition, and General Health *J. Nutr. Metab.* 2018: Article ID 7195760.

Chowdury, R., Chatterjee, R. Giri, A. K., Mandal, C. and Chauduri, K. (2010) Arsenic-induced cell proliferation is associated with enhanced ROS generation, Erk signaling and Cyclin A expression *Toxicol. Lett* 198: 263-71.

Circu, M. L. and Aw, T. Y. (2010) Reactive oxygen species, cellular redox systems and apoptosis *Free Radic. Biol. Med.* 48 (6): 749-62.

Cui, W., Luo, W. Zhou, X. Lu, Y., Xu, W., Zhong, S., Feng, G., Liang, Y., Liang, L. Mo, Y., Xiao, X. et al. (2019) Dysregulation of Ketone Body Metabolism Is Associated with Poor Prognosis for Clear Cell Renal Cell Carcinoma Patients *Front Oncol* 9: article 1422.

Daikhin, Y. and Yudkoff, M. (1998) Ketone bodies and brain glutamate and GABA metabolism *Dev Neurosci.* 20(4-5): 358-64.

D'Andrea-Meira, I., Romao, T. T., Jannuzzelli Pires do Prado, H., Kruger, L. T., Paiva Pires, M. E. and Oliveira do Conceicao, P. (2019) Ketogenic Diet and Epilepsy: What We Know So Far *Front Neurosci* doi: 10.3389/fnins.2019.00005.

Diano, S. and Horvath, T. L. (2012) Mitochondrial uncoupling protein 2 (UCP2) in glucose and lipid metabolism *Trends Mol. Med.* 18 (1): 52-58.

Digby, J. E., Martinez, F., Jefferson, A., Ruparelia, N., Chai, J., Wamil, M., Greaves, D. R. and Choudhury, R. P. (2012) Anti-inflammatory effects of nicotinic acid in human monocytes are mediated by

GPR109A dependent mechanisms *Arterioscler. Thromb vasc Biol* 32 (3): 669-76.

Dong, W., Jia, Y., Liu, X., Zhang, H., Li, T., Huang, W., Chen, X., Wang, F., Sun, W. and Wu, H. (2017) Sodium butyrate activates NRF2 to ameliorate diabetic nephropathy possibly via inhibition of HDAC *J. Endoc.* 232: 71–83.

Fine, E. J., Miller, A., Quadros, E. V., Sequeira, J. M. and Feinman, R. D. (2009) Acetoacetate reduces growth and ATP concentration in cancer cell lines which over-express uncoupling protein 2. *Cancer Cell Int.* 9:14.

Fleury, C., Neverova, M., Collins, S., Raimbault, S., Champigny, O., Levi-Meyrueis, C., Bouillaud, F., Seldin, M. F., Surwit, R. S., Ricquier, D. and Warden, C. F. et al. (1997) Uncoupling protein-2: a novel gene linked to obesity and hyperinsulinemia. *Nat Genet* 15: 269–272.

Franklin, S. T., Young, J. W. and Nonnecke, B. J. (1991) Effects of ketones, acetate, butyrate and glucose on bovine lymphocyte proliferation *J. Dairy Sci.* 74 (8): 2507-14.

Gibson, C. L., Murphy, A. N. and Murphy, S. P. (2012) Stroke outcome in the ketogenic state - a systematic review of the animal data *J Neurochem* 123 (2): 52-7.

Goldberg, E. L., Molony, R. D., Kudo, E., Sidorov, S., Kong, Y., Dixit, V. D. and Iwasaki, A. (2019) Ketogenic diet activates protective γδ T cell responses against influenza virus infection *Sci. Immunol.* 4 (41): 2026.

Han, Y-M., Bedarida, T., Ding, Y., Somba, B. K., Lu, Q. Wang, Q., Song, P. and Zou, M-H. (2018) B-Hydroxybutyrate Prevents Vascular Senescence through hnRNP A1-Mediated Upregulation of Oct4 *Mol Cell* 71: 1064-78.

Hartmann, H. L. and Vining, E. P. G. (2007) Clinical Aspects of the Ketogenic Diet *Epilepsia* 48 (1) 31-42.

Hasan-Olive, M. M., Lauritzen, K. H., Ali, M., Rasmussen, L. J., Storm-Mathisen, J. and Bergersen, L. H. (2019) A Ketogenic Diet Improves Mitochondrial Biogenesis and Bioenergetics via the PGC1α-SIRT3-UCP2 Axis *Neurochem. Res.* 44: 22–37.

Hole, P. S., Pearn, L., Tonks, A. J., James, P. E., Burnett, A. K., Darley, R.L. and Tonk, A. (2010) Ras-induced reactive oxygen species promote growth factor–independent proliferation in human CD34+ hematopoietic progenitor cells *Blood* 115: 1238-46.

Horton, W. E. & Sadler, T. W. (1983) Effects of maternal diabetes on early embryogenesis. Alterations in morphogenesis produced by the ketone body, B-hydroxybutyrate. *Diabetes* 37: 610–616.

Ji, F., Shen, T., Zou, W. and Jiao, J. (2017) UCP2 Regulates Embryonic Neurogenesis via ROS-Mediated Yap Alternation in the Developing Neocortex *Stem Cells* 35 (6): 1479-92.

Juge, N., Gray, J. A., Omote, H., Miyaji, T., Inoue, T., Hara, C., Uneyama, H., Edwards, R. H., Nicoll, R. A. and Moriyama, Y. (2010) Metabolic control of vesicular glutamate transport and release *Neuron* 68 (1): 99-112.

Kim, D. Y. and Rho, J. M. (2008) The ketogenic diet and epilepsy *Curr Opinion Clin Nutr Metab Care* 11 (2): 113-20.

Kimura, I., Inoue, D., Maeda, T., Hara, T., Ichimura, A., Miyauchi, S., Kobayashi, M. (2011) Short-chain fatty acids and ketones directly regulate sympathetic nervous system via G protein-coupled receptor 41 (GPR41) *Proc. Nat. Acad. Sci* 108 (19): 8030-5.

Koeslag, J. H., Noakes, T. D. and Sloan, A. W. (1980) Post-exercise ketosis *J Physiol* 301: 79-90.

Kong, G., Huang, Z., Ji, W., Wang, X., Liu, J., Wu, X., Huang, Z., Li, R. and Zhu, Q. (2017) The Ketone Metabolite β-Hydroxybutyrate Attenuates Oxidative Stress in Spinal Cord Injury by Suppression of Class I Histone Deacetylases *J. Neurotrauma* 34 (18) doi.org/10.1089/neu.2017.5192.

Kossof, E. H., Zupec-Kania, B. A., Amark, P. E., Ballaban-Gil, K. R., Bergqvist, A. G. C., et al. (2009) Optimal clinical management of children receiving the ketogenic diet: Recommendations of the International Ketogenic Diet Study Group *Epilepsia* 50 (2): 304-317.

LeBelle, J. E., Orozco, N. M., Paucar, A. A., Saxe, J. P., Mottahedeh, J., Pyle, A. D., Wu, H. and Kronblum, H. I. (2011) Proliferative Neural Stem Cells Have High Endogenous ROS Levels that Regulate Self-

Renewal and Neurogenesis in a PI3K/Akt-Dependent Manner *Cell Stem Cell.* 8(1): 59–71.

Ma, Q. (2013) Role of Nrf2 in Oxidative Stress and Toxicity *Annu Rev Pharmacol Toxicol.* 53: 401–426.

Ma, W., Berg, J. and Yellen, G. (2007) Ketogenic Diet Metabolites Reduce Firing in Central Neurons by Opening K_{ATP} Channels *J. Neurosci* 27 (14): 3618-3625.

Maalouf M., Rho J. M. and Mattson M. P. (2009). The neuroprotective properties of calorie restriction, the ketogenic diet, and ketone bodies. *Brain Res. Rev.* 59:293–315.

Madhavan L., Ourednik V. and Ourednik J. (2006) Increased "vigilance" of antioxidant mechanisms in neural stem cells potentiates their capability to resist oxidative stress. *Stem Cells.* 24(9):2110–9.

Marti, A., Larrarte, E. Novo, F. J., Garcia, M. and Martinez, J. A. (2001) UCP2 muscle gene transfer modifies mitochondrial membrane potential *Int. J. Obesity* 25: 68–74.

Miyamoto, J., Ohue-Kitano, R., Mukouyama, H., Nishida, A., Watanabe, K., Igarashi, M., Irie, J., Tsujimoto, G., Satoh-Asahara, N., Itoh, H. and Kimura, I. (2019) Ketone body receptor GPR43 regulates lipid metabolism under ketogenic conditions *Proc. Nat. Acad. Sci.* 116 (47): 23813-21.

Moore, D. C., Stanisstreet, M. and Clarke, C. A. Teratology (1989) Morphological and physiological effects of beta-hydroxybutyrate on rat embryos grown in vitro at different stages *Teratology* 40 (3): 237-51.

Musa-Veloso K., Likhodii S. S., Rarama E., Benoit S., Liu Y.-M., Chartrand C., Curtis D., Carmant R., Lortie L., Comeauc A., Cunnane F. J. E. (2006) Breath acetone predicts plasma ketone bodies in children with epilepsy on a ketogenic diet *Nutrition* 22(1):1–8.

Neal, E. G., Chaffe, H., Schwartz, R. H., Lawson, M. S., Edwards, N., Fitzsimmons, G., Whitney, A. and Cross, J. H. (2009) A randomized trial of classical and medium-chain triglyceride ketogenic diets in the treatment of childhood epilepsy *Epilepsia* 50 (5): 1105-17.

O'Donnell-Tormey J., Nathan C. F., Lanks K., DeBoer C. J., de la Harpe J. (1987) Secretion of pyruvate: an antioxidant defense of mammalian cells. *J. Exp. Med.* 165(2):500–514.

Pardo, A. C. (2020) Ketogenic diet: a role in immunity? *Paed. Neur. Briefs* doi.org/10.15844/pedneurbriefs-34-5.

Park, J. S. and Kim, Y. J. (2020) Anti-Aging Effect of the Ketone Metabolite β-Hydroxybutyrate in Drosophila Intestinal Stem Cells *Int. J. Mol. Sci.* 21: 3497-3510.

Pattappa G., Heywood H. K., De Bruijn J. D., Lee D. A. (2011) The metabolism of human mesenchymal stem cells during proliferation and differentiation. *J. Cell. Physiol.* 226:2562–2570.

Peterman, M. G. (1924) The ketogenic diet in the treatment of epilepsy *Am. J. Dis. Child.* 28 (1): 28-33.

Philips, M. C. L., Murtagh, D. K. J., Gilbertson, L. J., Asztely, F. J. S. and Lynch, C. D. P. (2018) Low-fat versus ketogenic diet in Parkinson's disease: A pilot randomized controlled trial *Mov Disord* 33 (8): 1306-14.

Puchalska,P. and Crwford, P. A. (2017) Multi-dimensional roles of ketone bodies in fuel metabolism, signaling, and therapeutics *Cell Metab.* 25(2): 262–284.

Quinlan, C. L., Perevoschchikova, I. V., Hey-Mogensen, M., Orr, A. L. and Brand, M. D. (2013) Sites of reactive oxygen species generation by mitochondria oxidizing different substrates *Redox Biol* 1(1):304-12.

Redza-Dutordoir, M. and Averill-Bates, A. (2016) Activation of apoptosis signalling pathways by reactive oxygen species *Biochim. Biophys. Acta* 1863 (12): 2977-92.

Reger, M. A., Henderson, S. T., Hale, C., Cholerton, B., Baker, L. D., Watson, G. S., Hyde, K., Chapman, D. and Craft, S. (2003) Effects of beta-hydroxybutyrate on cognition in memory-impaired adults *Neurobiol Aging.* 25(3):311-4.

Riquier, D. and Bouillaud, F. (2000) The uncoupling protein homologues: UCP1, UCP2, UCP3, StUCP and AtUCP *Biochem. J.* 345: 161–179.

Riechard, G. A.., Haff, A. C., Skutches, C. L., Paul, P., Holroyde, C. P. and Owen, O. E. (1979) Plasma acetone metabolism in the fasting human *J Clin Invest,* 63 (4): 619-26.

Robinson, A. M. and Williamson, D. H. (1980) Physiological roles of ketone bodies as substrates and signals in mammalian tissues *Physiol Rev* 60 (1): 143-87.

Robledinos-Anton, N., Fernandez-Gines, R., Manda, G. and Cuarado, A. (2019) Activators and Inhibitors of NRF2: A Review of Their Potential for Clinical Development *Oxid. Med. Cell. Long.* 2019: Article ID 9372182.

Rojas-Morales, P., Pedraz-Chaverri, J. and Tapia, E. (2020) Ketone bodies, stress response, and redox homeostasis *Redox Biol* 29: Article 101395.

Sato, S., Suzuki, T. and Okada, K. (1995) Suppression of mitogenic response of bovine peripheral blood lymphocytes by ketone bodies. *J Vet Med Sci* 57 (1) 183-5.

Shimazu, T., Hirschey, M. D., Newman, J., He, W., Shirakawa, K., Le Moan, N., Grueter, C. A., Lim, H., Saunders, L. R., Stevens, R. D., Newgard, C. B., Farese, R. V., de Cabo, R., Ulrich, S., Akassoglou, K. and Verdin, E. (2013) Suppression of Oxidative Stress by β-Hydroxybutyrate, an Endogenous Histone Deacetylase Inhibitor *Science* 339 (6116): 211–214.

Shukla, S. K., Gebregiworgis, T., Purohit, V., Chaika, N. V., Gunda, V., Radhakrishnan, P., et al. (2014) Metabolic reprogramming induced by ketone bodies diminishes pancreatic cancer cachexia. *Cancer Metab.* 2:18.

Simon, H-U., Haj-Yehia, A. and Levi-Schaffer, F. (2000) Role of reactive oxygen species (ROS) in apoptosis induction *Apoptosis* 5: 415–418.

Simsek T., Kocobas F., Zheng J., DeBarardinis R. J., Mahmoud A. I., Olson E. N., Schneider J. W., Zhang C. C., Sadek H. A. (2010) The distinct metabolic profile of hematopoietic stem cells reflects their location in a hypoxic niche. *Cell Stem Cell.* 7(3):380–390.

Skinner, R., Trujillo, A., Ma, X. and Beierle, E.A. (2009) Ketone bodies inhibit the viability of human neuroblastoma cells. *J Pediatr Surg.* 44:212–6.

Stephens, J. M., Sulway, M. J. and Watkins, P. J. (1971) Relationship of blood acetoacetate and 3-hydroxybutyrate in diabetes *Diabetes*, 20 (7): 485-9.

Sullivan, P. G., Rippy, N. A., Dorenbos, K., Concepcion, R. C., Agarwal, A. and Rho, J. M. (2004) The ketogenic diet increases mitochondrial uncoupling protein levels and activity *Ann Neurol.* 55(4):576-80.

Taggart, A. K. P., Kero, J., Gan, X., Cai, T. Q., Cheng, K., Ippolito, M., Ren, N., Kaplan, R., Wu, K., Wu, T-J., Jin, L., Liaw, C., Chen, R., Richman, J., Connolly, D., Offermans, S., Wright, S. and Waters, M. G. (2005) (D)-beta-Hydroxybutyrate inhibits adipocyte lipolysis via the nicotinic acid receptor PUMA-G *J Biol. Chem.* 280: 26649.

Vozza, A., Parisi, G., De Leonardis, F., Lasorsa, F. M., Castegna, A., Amorese, D., Marmo, R., Calcagnile, V. M., Palmieri, L., Ricquier, D., Paradies, E., Scarcia, P., Palmieri, F., Bouillaud, F. and Fiermonte, G. (2014) UCP2 transports C4 metabolites out of mitochondria, regulating glucose and glutamine oxidation *Proc Natl Acad Sci USA* 111(3): 960–965.

Wang, A. S., Ong, P. F., Chojnowski, A., Clavel, C. and Dreesen, O. (2017) Loss of lamin B1 is a biomarker to quantify cellular senescence in photoaged skin *Sci Reports* 7: article 15678. (a).

Wang, Q., Zhou, Y., Rychahou, P., Fan, T. W. M., Lane, A. N., Weiss, H. L. and Evers, M. (2017) Ketogenesis contributes to intestinal cell differentiation *Cell death and Diff.* 24 (3): 458-68. (b).

Weber, D. D., Aminazdeh-Gohari, S. and Kofler, B. (2018) Ketogenic diet in cancer therapy *Aging* 10(2): 164–165.

Wilder, R. M. (1921) High Fat diets in Epilepsy *Mayo Clin. Bull.* 2:307.

Wu, B. J., Yan, L., Charlton, F., Witting, P., Barter, P. J. and Rye, K-A. (2010) Evidence That Niacin Inhibits Acute Vascular Inflammation and Improves Endothelial Dysfunction Independent of Changes in Plasma Lipids *Arterioscler. Thromb. Vasc. Biol.* 30: 968-75.

Xie, Z., Zhang, D., Chung, D., Tang, Z., Huang, H., Dai, L., Qi, S., Li, J., Colak, G., Chen, Y., Xia, C., Peng, C., et al. (2016) Metabolic Regulation of Gene Expression by Histone Lysine β-Hydroxybutyrylation *Mol. Cell* 62 (2) 194-206.

Zhang J., Khvorostov I., Hong J. S., Oktay Y., Vergnes L., Neubel E., Wahjudi P. N., Setoquchi K., Wang G., Do A., Jung H. J., McCaffery J. M., Kurland I. J., Reue K., Lee W. N., Koehler C. M., Teitell M. A. (2011) UCP2 regulates energy metabolism and differentiation potential of human pluripotent stem cells. *EMBO J.* 30(24):4860–4873.

Zhang, Q., Wang, S. P., Li, L. L. and Zhang, Y. (2017) Effects of mitochondrial uncoupling protein 2 inhibition by Genipin on rat bone marrow mesenchymal stem cells under hypoxia and serum deprivation (H/SD) conditions *Int J Clin Exp Pathol* 10(9):10047-10055.

Zhang, F. Q., Jiang, J. L., Zhang, J. T., Niu, H., Fu, X. Q. and Zeng, L. L. (2020) Current status and future prospects of stem cell therapy in Alzheimer's disease *Neural Regen. Res.* 15 (2): 242-50.

BIOGRAPHICAL SKETCH

Mary Board

Affiliation: St. Hilda's College, University of Oxford, UK.

Education: MA, DPhil (Oxon)

Research and Professional Experience: Lecturer and Tutor in Biochemistry, St. Hilda's College, University of Oxford; University Research Fellow.

Publications from the Last 3 Years:

Key publications:

1. Faulkner, A. R. M., Lynam, E., Purcell, R., Jones, C., Lopez, C., Board, M., Wagner, K., Wagner, N., Carr, C. & Wheeler-Jones, C., (2020) *Context-dependent regulation of endothelial cell*

metabolism: differential effects of the PPARβ/δ agonist GW0742 and VEGF-A.
2. *Scientific Reports.* 10, 1, 15 p., 7849.
3. Chen, T., Hua, Y., Li, Z., Urban, J., Board, M., Cui, Z. and Huidong, J. (2020) Improving transport and storage of viable mesenchymal stem cells through investigations into their energy metabolism. *London cell Cycle Club*, Francis Crick institute.
4. Chen, T., Hua, Y., Li, Z., Urban, J., Board, M., Cui, Z. and Huidong, J. (2020) *Improving transport and storage of viable mesenchymal stem cells through investigations into their energy metabolism.* Tissue Engineering and Regenerative Medicine International Society (TERMIS) (European Chapter).
5. Board, M., Lopez, C., van den Bos, C., Callaghan, R., Clarke, K. and Carr, C. (2017) Acetoacetate is a more efficient energy-yielding substrate for human mesenchymal stem cells than glucose and generates fewer reactive oxygen species. *Int. J. Biochem Cell Biol.* 88, 75-83
6. Two Chapters "Carbohydrate Metabolism," Board, M. and "Fat Metabolism," Board, M. (2017) in synoptic guide for Membership of Royal College of Obstetricians and Gynaecologists, CUP.
7. Smith, H., Board, M., Pellagatti, A., Turley, H, Boultwood, J., and Callaghan, R. (2016) The effects of severe hypoxia on glycolytic flux and enzyme activity in a model of solid tumours *J. Cell. Biochem.* 1999, 1-12.

INDEX

A

acid, 73, 74, 104, 106, 107, 110, 119, 125
acute respiratory distress syndrome, 7, 12
adipose tissue, 13, 25, 57, 110
airway epithelial cells, 6
alveoli, 5, 8, 9, 16
angiogenesis, 14, 18, 58, 62, 66, 67, 70, 72, 74, 75, 76
antibody, 18, 21, 23
anti-cancer, ix, 101, 114
antigen-presenting cells (APCs), 18
anti-inflammatory agents, 29
antioxidant, 108, 122, 123
apoptosis, ix, 8, 58, 66, 71, 76, 102, 103, 108, 110, 113, 115, 116, 117, 119, 123, 124
arterial hypertension, 23
autoimmune diseases, 21, 24
autosomal recessive, 23

B

beneficial effect, 21, 64, 65, 70, 74, 76
benefits, vii, x, 3, 21, 59, 102
biomaterials, 63
bipolar disorder, 77
bleeding, 55
blood, 5, 7, 11, 16, 22, 31, 33, 42, 45, 55, 68, 69, 72, 93, 105, 106, 115, 124, 125
blood flow, 55, 72
blood pressure, 5
blood supply, 55
blood vessels, 11
blood-brain barrier, 68
bloodstream, 105
bone marrow, 13, 57, 58, 69, 72, 73, 126
brain, 5, 21, 27, 57, 58, 61, 62, 63, 64, 65, 67, 68, 69, 70, 72, 74, 75, 76, 77, 84, 89, 105, 119
brain stem, 27
breast cancer, 114, 118

C

caloric restriction, ix, 101
cancer, x, 17, 21, 28, 102, 113, 116, 118, 120, 124, 125

cancer cells, 17, 118
cancer stem cells, 116
cancer therapy, x, 102, 125
carbohydrate, ix, 101, 103, 105, 107, 112, 115
carbohydrate metabolism, ix, 101, 103, 105
carbon, 104, 106, 110
cardiac arrhythmia, 25
cardiovascular disease, 76
cell culture, 69, 117
cell cycle, ix, 18, 102, 108, 113, 114
cell death, 8, 17, 58, 66, 71, 75
cell differentiation, 25, 117, 125
cell division, 103
cell line, 76, 120
central nervous system, 105
chemicals, 60, 62, 73, 74, 75
chemokine receptor, 14, 54, 68
chemokines, 27, 63, 64
China, 4, 5, 7, 31, 32, 40, 48, 51, 76
chronic obstructive pulmonary disease, 23
clinical trials, vii, viii, 3, 20, 21, 22, 23, 24, 26, 27, 28, 30, 53, 56, 57, 64
coagulopathy, viii, 2, 11, 12, 13, 29, 30
colorectal cancer, 113
conditioning, 57, 62, 66
consumption, ix, 102, 106, 109
cultural conditions, 69
culture, vii, x, 16, 57, 60, 69, 78, 102, 103, 112, 116
culture conditions, 60
culture medium, 103, 117
cystic fibrosis, 23
cytokines, vii, ix, 7, 9, 16, 17, 18, 20, 21, 23, 24, 27, 54, 60, 61, 62, 63, 64, 65, 78
cytoplasm, 10

D

D-3-hydroxybutyrate, ix, 101, 102, 107
deaths, viii, 2, 19, 29

degradation, 7, 18
dendritic cell, 16, 18, 19, 24
deprivation, 60, 126
detection, 7, 8, 20, 27
diabetes, 118, 121, 125
diabetic ketoacidosis, 105
diabetic nephropathy, 120
diet, vii, ix, 101, 104, 115, 120, 121, 122, 123, 125
diseases, vii, viii, 2, 3, 16, 21, 23, 28, 30, 76
disseminated intravascular coagulation, 11
drugs, vii, ix, 28, 54, 60, 62, 63, 73, 75

E

edema, 9, 10, 22, 68
embolism, 12, 29
endothelial cells, viii, 2, 5, 11, 12, 16, 25, 29, 65, 68
energy, ix, 101, 103, 105, 106, 111, 113, 118, 126, 127
energy expenditure, 106, 111
environment, viii, 29, 54, 59, 109
enzyme, 5, 10, 12, 127
epigenetic modification, 63
epilepsy, x, 102, 104, 115, 121, 122, 123
epithelial cells, ix, 6, 101, 105, 112, 113
evidence, vii, viii, 27, 53, 58, 103, 107, 109, 114, 116
exercise, 111, 118, 121
exposure, 63, 108, 113

F

fasting, 112, 118, 119, 124
fat, ix, 101, 103, 105, 115, 123
fatty acids, 105, 106, 121
fibroblast growth factor, 14
formation, 4, 6, 11, 71, 73

Index

G

gene expression, ix, 102, 112, 114
gene transfer, 122
genes, ix, 17, 24, 25, 70, 72, 102, 109, 112, 114
glucose, ix, 101, 103, 105, 107, 109, 110, 118, 119, 120, 125, 127
glutamate, 115, 119, 121
growth, vii, ix, 14, 54, 55, 60, 62, 64, 65, 66, 74, 114, 120, 121
growth factor, vii, ix, 14, 54, 55, 60, 62, 64, 66, 74, 121

H

HB, vii, ix, 101, 103, 104, 106, 109, 111, 112, 113, 114, 115, 116, 117
health condition, 77
hematopoietic stem cells, 72, 124
histone, ix, 77, 102, 112
histone deacetylase, ix, 77, 102, 112
host, 4, 6, 14, 15, 16
human, ix, 4, 25, 57, 59, 64, 70, 72, 92, 102, 103, 104, 109, 116, 118, 119, 121, 123, 124, 126, 127
hydrogen peroxide, 108
hydroxyl, 108
hyperinsulinemia, 120
hypertension, 23
hypothesis, 30
hypoxemia, 22, 25
hypoxia, vii, ix, 8, 54, 60, 62, 69, 71, 72, 74, 76, 78, 126, 127

I

immune disorders, viii, 2
immune modulation, 15
immune reaction, 24
immune response, 9, 10, 12, 19, 65, 69, 113
immune system, viii, 2, 3, 6, 8, 9, 16, 17, 21, 23, 28, 56
immunomodulatory, 15, 16, 19, 56, 65
immunosuppression, 24, 27, 62
in vitro, vii, x, 13, 17, 25, 57, 61, 64, 69, 71, 77, 83, 84, 93, 99, 102, 103, 109, 114, 116, 122
in vivo, vii, viii, x, 13, 53, 61, 64, 65, 71, 73, 79, 83, 84, 99, 102, 103, 104, 106, 117
infarction, 59, 72, 74, 75, 102
infection, viii, 2, 8, 9, 12, 13, 15, 16, 23, 30, 113, 120
inflammation, 3, 8, 9, 10, 11, 13, 14, 15, 16, 21, 22, 23, 27, 29, 30, 58, 64, 65
inflammatory mediators, 19, 65
inflammatory responses, 64, 70
influenza, 2, 25, 28, 120
influenza virus, 25, 120
ingredients, 29
inhibition, 19, 77, 112, 113, 120, 126
inhibitor, 5, 14, 74, 110, 114
injury, iv, 25, 59, 64, 69, 76
integrity, 4, 10, 113
irritable bowel syndrome, 113
ischemia, 11, 55, 59, 65, 74, 76

K

ketoacidosis, 105, 117
ketone bodies, v, vii, ix, 101, 102, 103, 104, 107, 111, 112, 113, 118, 119, 122, 123, 124
ketones, 113, 120, 121
kidney, 3, 5, 13, 21, 109
kidney failure, 13
killer cells, 72

L

laboratory studies, 57
lipid metabolism, 119, 122
lipids, 108
lipolysis, 111, 125
liver, ix, 3, 13, 57, 101, 105
liver failure, 13
lung function, 22, 25
lymphocytes, 17, 19, 112, 118, 124

M

machinery, 73, 108, 110
macromolecules, 108
macrophages, 16, 17, 18, 19, 20, 23
malignant cells, 114
mammalian cells, 106, 123
mammalian tissues, 124
management, 73, 76, 121
marrow, 13, 54, 57, 58, 69, 71, 72, 73, 126
matrix metalloproteinase, 72
mesenchymal stem cells, v, vii, viii, ix, 1, 2, 3, 13, 31, 32, 33, 34, 36, 38, 39, 40, 41, 42, 43, 44, 46, 47, 48, 50, 51, 53, 54, 56, 71, 78, 79, 80, 81, 82, 84, 85, 86, 87, 88, 89, 90, 91, 92, 93, 94, 95, 96, 97, 98, 99, 102, 103, 104, 107, 110, 118, 123, 126, 127
metabolic change, 103, 117
metabolic changes, 103, 117
metabolism, ix, 79, 80, 81, 91, 93, 96, 98, 101, 102, 104, 106, 107, 108, 109, 111, 114, 118, 119, 122, 123, 124, 126, 127
migration, 8, 9, 14, 16, 18, 59, 61, 63, 68, 69, 70, 71, 74, 76, 77
mitochondria, 123, 125
models, vii, viii, 23, 25, 53, 68, 74, 116
molecular weight, 13
molecules, 4, 6, 14, 15, 17, 18, 19, 20, 28, 61
monoclonal antibody, 21
monocyte chemoattractant protein, 9
multiple sclerosis, 16
multipotent, vii, viii, 13, 53, 69, 92
muscle mass, 106
muscles, 106
mutation, ix, 102, 117
mutations, 5, 23, 103, 108
myocardial infarction, 72, 102

N

National Academy of Sciences, 33, 85
National Institutes of Health, 57
nerve growth factor, 54
nervous system, 3, 58, 105, 111, 121
neuroblastoma, 114, 124
neuroblasts, 67
neurodegeneration, 117
neurodegenerative disorders, x, 102, 104, 115, 118
neurogenesis, 58, 67, 70, 74, 77
neuroprotection, 65, 70, 114
neuroscience, 91
neurotransmitter, 115
nicotinic acid, 119, 125
NK cells, 8, 17, 18, 19

O

organ, viii, 2, 8, 9, 11, 12, 13, 21, 24, 27, 28, 30
organs, viii, 2, 5, 12, 13, 21, 23, 28, 29, 30, 57, 69
oxidation, 103, 106, 107, 110, 113, 125
oxidative damage, ix, 102
oxidative stress, ix, 58, 61, 102, 108, 109, 115, 117, 122

P

pancreatic cancer, 114, 124
pathogenesis, 12, 20
pathogens, 10, 18
pathology, viii, 54, 67
pathophysiology, 64
pathway, 17, 20, 75, 77, 105, 107, 108
peripheral blood, 69, 124
phagocytic cells, 9, 16
pharmaceutical, 30
pharmacological agents, 60, 73, 74
pharmacological treatment, viii, 2
phenotype, 18, 19, 20, 23, 61, 65, 118
pneumonia, viii, 2, 3, 8, 22, 26
population, 17, 72, 103, 108, 109, 113, 114, 117
positive feedback, 20
post-transplant, viii, 54, 78
potential benefits, vii, x, 102
priming, viii, 54, 57, 60, 61, 62, 63, 65, 66, 67, 68, 71, 73, 74, 75, 77, 78
pro-inflammatory, 16, 18, 20, 28, 29, 65, 71, 78
proliferation, ix, 14, 15, 16, 17, 18, 19, 24, 25, 67, 69, 71, 73, 76, 102, 104, 108, 109, 112, 113, 114, 116, 117, 118, 119, 120, 121, 123
prostaglandin, 14, 65
protection, 57, 60, 110
proteins, ix, 3, 6, 11, 15, 28, 102, 105, 112, 114
prothrombin time, 11
pulmonary edema, 22
pulmonary embolism, 29

oxygen, ix, 18, 26, 61, 69, 72, 78, 102, 103, 107, 108, 118, 119, 121, 123, 124, 127

R

reactive oxygen, ix, 18, 61, 102, 103, 107, 118, 121, 123, 124, 127
reading, 4
receptor, 4, 5, 10, 12, 14, 15, 19, 20, 23, 54, 68, 77, 112, 121, 122, 125
receptors, viii, ix, 2, 5, 6, 8, 10, 11, 15, 101, 105
recommendations, iv
reconditioning, 63, 64, 72, 75
recovery, vii, viii, 53, 56, 59, 65, 66, 67, 68, 70, 72, 74, 76, 77
regenerate, 25, 113
regeneration, 8, 18, 25, 27, 58, 63, 68, 70, 73, 77, 103, 113, 117
regenerative capacity, 113
regenerative medicine, 56
repair, vii, viii, 15, 21, 25, 27, 53, 64, 103, 117
replication, 7, 8, 12, 28, 29, 76
requirement, 104, 106
respiratory distress syndrome, 7, 12, 22
respiratory syncytial virus, 28
response, 8, 9, 10, 12, 17, 18, 19, 20, 27, 28, 29, 56, 59, 64, 65, 109, 112, 113, 119, 124
risk, 29, 55, 73, 117
rodents, 66, 71, 72, 74

S

SARS, viii, 2, 3, 4, 5, 7, 10, 11, 16, 25, 28, 29, 35, 36, 37, 39, 45, 48
SARS-CoV, viii, 2, 3, 4, 5, 10, 11, 16, 25, 28, 29, 35, 36, 37, 39, 45, 48
secrete, 8, 14, 18, 29, 68
secretion, 8, 17, 18, 19, 65, 71
serum, 60, 67, 77, 126
shortness of breath, 7, 8, 9, 26

signal transduction, 29
signaling pathway, 75, 77
signalling, 103, 108, 111, 113, 114, 123
signals, 10, 25, 70, 124
significant, 23, 56, 64, 66, 72, 76, 102, 103, 104, 108, 111
small intestine, 5
smooth muscle, 5
smooth muscle cells, 5
species, ix, 18, 61, 102, 103, 107, 118, 119, 121, 123, 124, 127
starvation, ix, 101, 105, 111, 114, 119
stem cell differentiation, 117
stem cells, vii, viii, ix, 3, 6, 8, 25, 29, 53, 54, 56, 58, 67, 68, 69, 71, 72, 102, 103, 104, 107, 108, 109, 110, 112, 113, 114, 116, 118, 122, 123, 124, 126, 127
stimulation, 19, 103, 111
stress, ix, 58, 61, 102, 108, 109, 116, 117, 122, 124
stress response, 124
stroke, vii, viii, 13, 53, 55, 56, 58, 62, 63, 64, 67, 68, 69, 71, 72, 73, 74, 75, 76, 98, 104, 116
stromal cells, 13, 92
structural protein, 4
structure, 4, 5, 6, 9, 16
substrates, ix, 72, 101, 103, 104, 105, 111, 113, 114, 123, 124
survival, viii, 12, 23, 24, 53, 60, 61, 66, 70, 71, 74, 75
survival rate, 24, 74
sympathetic nervous system, 58, 111, 121
symptoms, vii, viii, 2, 3, 5, 7, 8, 12, 13, 21, 22, 23, 26, 28, 29, 30
synaptic vesicles, 115
syndrome, viii, 2, 9, 19, 22, 24, 73, 113

T

T cell, 8, 16, 17, 18, 19, 120

target, 5, 11, 13, 15, 28, 56, 59, 64, 103, 105, 109, 111
techniques, 60, 64, 78
therapeutic approaches, 56
therapeutic effect, 59, 60, 66
therapeutic use, 14, 111
therapeutics, 24, 123
therapy, x, 24, 27, 29, 56, 58, 69, 98, 102, 104, 111, 114, 116, 117, 125, 126
tissue, 14, 15, 18, 25, 27, 55, 56, 57, 59, 64, 70, 72, 77, 102, 105, 110, 115, 117
TNF, 16, 17, 18, 31, 55, 64, 65, 72
TNF-α, 16, 17, 18, 31, 55, 65, 72
transcription, 7, 69, 108, 109, 113, 114
transplantation, vii, viii, x, 24, 26, 54, 56, 59, 60, 61, 62, 69, 73, 75, 78, 102, 104, 115, 116
transport, 4, 15, 59, 108, 110, 121, 127
treatment, vii, viii, 2, 5, 13, 16, 21, 24, 25, 26, 27, 28, 29, 30, 53, 56, 58, 59, 60, 62, 63, 64, 73, 75, 76, 104, 115, 122, 123
trial, vii, viii, 2, 3, 21, 23, 25, 26, 30, 102, 122, 123
tricarboxylic acid, 106, 107, 110
tricarboxylic acid cycle, 106, 107, 110
tumor necrosis factor, 9, 65
tumour suppressor genes, 114

U

umbilical cord, 13, 57, 72
upper respiratory tract, 5

V

ventricular tachycardia, 25
viral infection, viii, 2, 8, 30
virus infection, 120
viruses, 3, 6, 7, 8, 9, 10, 11, 28